General Science

BOOK TWO

General Science

BOOK TWO

by Charles Windridge

edited by Patrick J. Kenway, B.Sc.

illustrated by Barry Davies

SCHOFIELD & SIMS LTD · HUDDERSFIELD

0 7217 3552 5
0 7217 3557 6   Net edition

First edition 1958
Reprinted 1959 (thrice), 1960, 1961, 1962 (twice), 1963,
1964, 1965, 1966, 1967, 1968
Second (revised) edition 1969
Reprinted 1970, 1971, 1972, 1973
Third (revised) edition 1976
Reprinted 1977, 1978

GENERAL SCIENCE is a series of five books.

Book One   [0 7217 3551 7]
Book Two   [0 7217 3552 5]
Book Three [0 7217 3553 3]
Book Four [0 7217 3554 1]
Book Five [0 7217 3555 X]

Printed in England by
Henry Garnett & Co. Ltd., Rotherham and London

Author's Note

REVISED EDITION

This book is a completely up-dated and fully revised edition of General Science, Book Two, which was first published in 1958.

The complete revision of this book and the other books in the series ensures that they are fresh in their approach and that the series as a whole will satisfy the needs of pupils in this day and age—the Space Age. The revision has been carried out with a proper regard for the latest developments in scientific research and thought, the recent trends in the methodology of science teaching, and the ever-changing pattern of technological advancement, and in the awareness that science is not just a body of knowledge but also a set of methods and a way of thinking. Much of the revision has been concerned with those modifications that have been made necessary by the omission of obsolescent material, the introduction of SI metric units, new discoveries, changes in examination requirements, etc.; but the series still retains those familiar characteristics which, as very many reprints over the years clearly indicate, have made it very popular and established it as being thoroughly reliable and adequate.

This series provides a course in general and combined science that is complete, but in no way overloaded, for pupils of average ability within the 11 to 16 age-range. It is essentially a study of natural science—physics, chemistry, biology, astronomy, geology, meteorology, etc.—and its applications in technology and other fields. The fifth book in this series caters for those pupils who are taking preliminary technical courses or examinations in general science of the standard of the Certificate of Secondary Education, and it provides a sound groundwork for the important minority who will take trade courses and advanced courses in technology later on.

The course is presented as an integrated and solid core of indispensable and basic facts and techniques combined with coherent topics that will promote purposeful activities, encourage individual research and satisfy modern examination requirements. There are opportunities for pupils to acquire and use knowledge and skills in both real and realistic situations. They are made to understand that the ability to handle scientific situations, rather than the mere mental storage of scientific facts, is the proper end-product of scientific education.

An important feature of the books in this series is that both the diagrams and the text are almost entirely self-explanatory, and this, coupled with the tremendous variety and the wide coverage in the contents, means that most pupils could work from them without very much assistance from the teacher. Therefore, in addition to their usefulness as textbooks of the traditional type, the books are very suitable for use with mixed ability classes and for homework, projects and individual study.

The practical work is straightforward and homely, so that, quite often, it can be performed with simple or improvised apparatus. The diagrams are examples on which pupils can base their own drawings. The exercises, given at the end of the books, are intended to supplement the pupils' practical work and records of experiments by providing extra practice in written work, drawing, the use of scientific terms, calculations, etc., and, of course, they are useful for revision purposes. The units of measurement are S I (Le Système International d' Unités) and decimal notation is used almost exclusively throughout.

Full consideration has been given to the various reports and recommendations of such bodies as the Certificate of Secondary Education and University Examination Boards, the Association for Science Education, the Council of Technical Examining Bodies, the Royal Society, the Royal Institute of Chemistry, etc.

C.W.

Metric Units

THE STANDARD INTERNATIONAL SYSTEM OF UNITS

Le Système International d' Unités, which is known here as the Standard International System of Units, or, more simply, as S I, is the official measuring system of the United Kingdom, and all the units of measurement that are used in this series of books are part of this system except in a few instances where, for historical or some other special reasons, it is necessary to do otherwise.

Some of the S I fundamental, derived and supplementary units are:

physical quantity	unit	symbol
length	metre	m
mass	kilogram	kg
time	second	s
temperature	kelvin	K
temperature	degree Celsius (customary unit)	°C
electric current	ampere	A
luminous intensity	candela	cd
area	square metre	m^2
volume	cubic metre	m^3
velocity	metre per second	m/s
acceleration	metre per second per second	m/s^2
force	newton	N
heat, work and energy	joule	J
power	watt	W

The above-mentioned units and some others not already mentioned are defined and fully explained at appropriate places in the text.

As far as is possible all fractions are expressed in decimal notation so that full benefit is derived from a system of units whose multiples and sub-multiples can be so easily expressed in that same notation.

Further information about S I units and conventions can be obtained from the publications that are issued from time to time by such bodies as the Royal Society, the Association for Science Education, the British Standards Institution, the Royal Institute of Chemistry, Her Majesty's Stationery Office, the Council of Technical Examining Bodies, the Department of Education and Science, etc. The booklets *The Use of S I Units in the Early and Middle Years of Schooling* and *S I Units, Signs, Symbols and Abbreviations for Use in School Science*, both published by the Association for Science Education, are particularly helpful publications.

Contents

Fruits and Seeds

Some common fruits

Examine and make labelled drawings of the fruits of some common plants. Some of these are shown in the illustrations opposite. Notice the remains of flowers and stems. Break open some of these fruits and examine the seeds inside them. Notice their shapes, sizes and colours.

Collecting fruits and seeds

Collect some fruits and seeds of wild and garden plants. Store them in test-tubes and match-boxes or on display cards. Attach labels. Small juicy fruits likely to decay rapidly can be preserved cheaply by dipping them in molten paraffin wax. The molten wax must not be too hot or the specimens will be destroyed.

Do *not* eat any of the fruits and seeds you collect. Some fruits and seeds are poisonous.

Seed dispersal

It would not do for all seeds to fall and grow beneath their parent plants. The result would be overcrowding. Seeds are carried away from their parent plants by *wind, water, birds, mammals* and the *special mechanisms* which some plants possess.

Ash and sycamore fruits have flat wings. The dandelion, clematis and thistle have hairy growths which allow them to be carried, parachute-fashion, by the wind. Have you not heard the saying "as light as thistledown"? Wind causes the ripe fruits of the poppy to sway, and its seeds are thrown out. Seeds which are very light in weight are also carried by the wind. Orchid seeds are almost as fine as dust.

Fruits which float, such as those of the water-lily and the coconut palm, are carried by water. Coconuts travel for thousands of miles across seas and oceans. The original coconut palms on the islands in the South Pacific Ocean grew from fruits which were carried there from the mainland by ocean currents.

The seeds of juicy fruits are sometimes carried by birds and other animals. The fruits are eaten but only the juicy parts are digested. The undigested stones and pips leave the bodies of the animals at places which may be many miles away from the parent plants. Blackberry, cherry and apple seeds are dispersed in this way.

Small seeds are carried in the mud on the feet of birds and other animals. Fruits with hooks and prickles, such as those of the burdock, are carried in the hair, fur and wool of animals.

Some plants disperse their own seeds. Their pods dry and then split open suddenly with a force which throws their seeds for some distance. The laburnum and vetch disperse their seeds in this way.

Some seeds which are dispersed

Examine and make labelled drawings of any of the following fruits which are available. State the method of seed dispersal used.

Sycamore; ash; clematis; dandelion; thistle; water-lily; water-plantain; onion; poppy; rose; apple; cherry; burdock; laburnum; vetch; pea; horse-chestnut; goose-grass.

Growing seeds

Try growing some of the seeds you have collected. Soak them in water for at least twenty-four hours. Then sow them about 5 mm deep in pots containing a mixture that is half of sand and half of *loam*. Loam is a rich soil of sand, clay and vegetable matter. Stand the pots in a warm place and keep the mixture moist.

Sow bird seeds, of the kind used for feeding canaries, pigeons and budgerigars, in a dish of soil. Keep them warm and moist. These seeds germinate and grow very quickly. They belong to the grass and the pea and bean families of plants.

Grow an acorn in the neck of a medicine bottle filled with water. You will be able to see the roots growing down into the water.

SOME COMMON FRUITS

(not to scale)

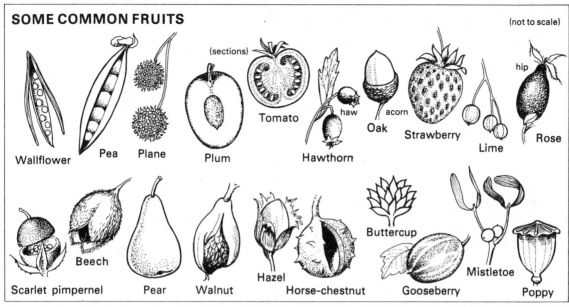

Wallflower · Pea · Plane · Plum · Tomato (sections) · Hawthorn · haw · acorn · Oak · Strawberry · Lime · hip · Rose

Scarlet pimpernel · Beech · Pear · Walnut · Hazel · Horse-chestnut · Buttercup · Gooseberry · Mistletoe · Poppy

COLLECTING FRUITS AND SEEDS

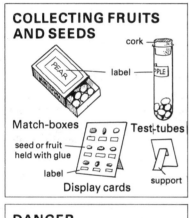

cork
label
Match-boxes
PEAR
APPLE
Test-tubes
seed or fruit held with glue
label
support
Display cards

DANGER

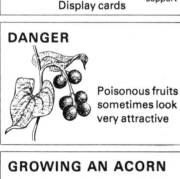

Poisonous fruits sometimes look very attractive

GROWING AN ACORN

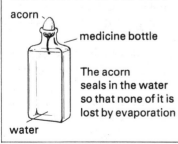

acorn
medicine bottle
The acorn seals in the water so that none of it is lost by evaporation
water

SEED DISPERSAL

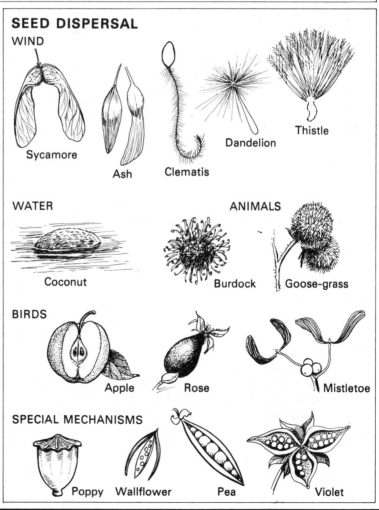

WIND

Sycamore · Ash · Clematis · Dandelion · Thistle

WATER

Coconut

ANIMALS

Burdock · Goose-grass

BIRDS

Apple · Rose · Mistletoe

SPECIAL MECHANISMS

Poppy · Wallflower · Pea · Violet

2   *Vegetative Reproduction*

Vegetative reproduction

Plants usually grow from seeds. Some plants, however, reproduce themselves by means of special kinds of organs which are called *runners, rhizomes, tubers, bulbs* and *corms*. This process is called *vegetative reproduction*.

Runners and rhizomes

Runners are creeping stems which grow out of a parent plant. Roots and buds grow on these runners and new plants develop. The parent plant supplies food to the young plants for a time, and then the runners die and the young plants grow independently. Strawberry runners grow above the ground and buttercup runners grow below the ground.

Rhizomes are long underground stems. Roots and leaves grow at intervals along them. Couch grass has a rhizome. Some rhizomes, such as those of Solomon's seal, are swollen and contain food.

Tubers

Tubers are swollen roots or underground stems. Potatoes are swollen stems. Below the ground, a potato is not green, because it contains no *chlorophyll*. A potato growing above the ground is green because sunlight causes the formation of chlorophyll. The *eyes* of a potato are small buds. When a potato is planted in soil, it sprouts and its buds grow into new potato plants. Gardeners grow potatoes from seeds—a difficult business—when new varieties or disease-free plants are required. *Note: seed potatoes* are not seeds but tubers.

Bulbs

Bulbs are large underground buds with swollen food leaves. A bulb contains food leaves, *scale leaves*, a short stem, next year's buds attached to this stem, and fibrous roots. A bulb contains, conveniently packed, all the organs of a flowering plant. Plants which are grown from bulbs can often be grown from seeds, but gardeners prefer to use bulbs because the food contained in their leaves gives the young plants such a good start in life. The onion, snowdrop, shallot, daffodil, scilla, hyacinth and tulip are bulb-plants. There are many more.

Corms

Corms are often mistaken for bulbs. A corm is a flat swollen stem filled with food. It contains no leaves apart from the protective scale leaves on its outside and those in the shoots. A corm contains next year's buds, a flat swollen stem, the remains of last year's corm and fibrous roots. At the beginning of the season, the buds grow and the shoots use the food contained in the swollen stem. At the end of the season, the plant dies down and new corms form at its base. These corms contain food for the next year's plants. Crocuses, anemones and gladioli have corms.

Some organs of vegetative reproduction

Examine and make labelled drawings of a potato tuber and *longitudinal sections* of an onion bulb and a crocus corm. Notice the following:
Potato: remains of stem, *lenticels* (breathing pores), swollen tuber and buds.
Onion: scale leaves, food leaves, next year's buds, stem and roots.
Crocus: scale leaves, buds, swollen food stem, last year's corm and roots.

Some bulbs and corms

Examine and make labelled drawings of some common bulbs and corms.

Food stores in plants

Grow some potato tubers and onion bulbs on jam-jars filled with water. Cover the jars with paper funnels; the growth of shoots begins best in darkness.

The tubers and the bulbs become soft and decrease in size. The food stores in them are being used up by the growing plants.

Remove the shoots from one of the potatoes. What happens to it? Does it continue to grow?

VEGETATIVE REPRODUCTION

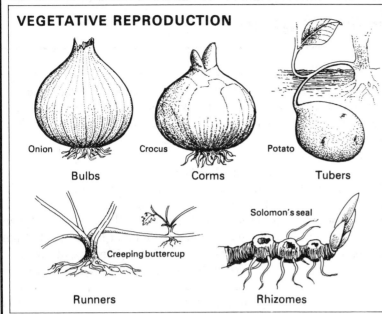

Onion — Crocus — Potato

Bulbs   **Corms**   **Tubers**

Creeping buttercup

Solomon's seal

Runners   **Rhizomes**

FOOD STORES IN PLANTS

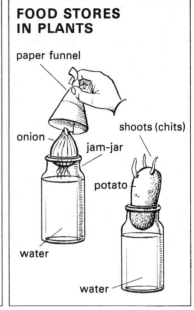

paper funnel

shoots (chits)

onion

jam-jar

potato

water

water

POTATO TUBER

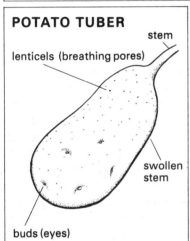

stem

lenticels (breathing pores)

swollen stem

buds (eyes)

ONION BULB   (section)

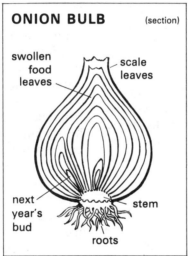

swollen food leaves

scale leaves

next year's bud

stem

roots

CROCUS CORM   (section)

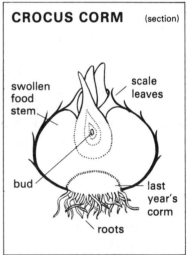

swollen food stem

scale leaves

bud

last year's corm

roots

SOME BULBS AND CORMS

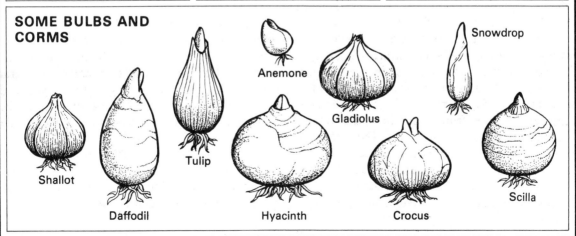

Anemone

Snowdrop

Gladiolus

Shallot

Tulip

Daffodil

Hyacinth

Crocus

Scilla

Preparing for Winter

Preparing for winter

Plants do not grow at the same rate throughout the year. Plant growth is at a maximum during spring. Plants transpire and breathe more in summer than in winter. During spring, leaves and buds are forming; the sap is rising up the stems and the plants are increasing in size.

In autumn, herbaceous plants die down and deciduous trees lose their leaves. They are preparing for winter. Trees are not active during winter. There is no growth, and breathing and transpiration are just sufficient to keep them alive.

Leaf fall

Leaf fall does not take place suddenly. During late summer, trees grow layers of cork between their twigs and the ends of their leaf stalks. In autumn, the sap does not reach the leaves and they wither and fall. *Vein endings* can be seen on the stalk of a fallen leaf. The *leaf scars* on twigs indicate the places from which the leaves have fallen. The veins are sealed off by corky tissue. Autumn leaves take on various shades of yellow, red, brown and purple. The fallen leaves rot to form *leaf-mould,* which is rich in food for plants.

Seasonal changes in growth

The seasonal changes in the rate of growth of a tree are shown by the cut end of its trunk. The *annual rings,* each of which is one year's growth, can be counted and the age of the tree determined. These rings are formed of alternate layers of wood of different textures. The wood formed by rapidly growing trees in spring is not so dense and so dark in colour as that formed in summer and early autumn. The hard wood in the centre of a tree trunk contains little or no moisture. It is called *heartwood.* The soft, sap-filled wood near to the bark is called *sapwood.* The *knots* in timber indicate the positions of branches. One of the illustrations shows how the heartwood continues from a trunk into a branch.

Annual rings

Examine and make a labelled drawing of the cut end of a tree trunk (*transverse section*). Notice the annual rings, heartwood, sapwood and bark. How old was the tree when it was felled?

Growing miniature trees

Collect seeds of the sycamore, laburnum, apple, hazel, oak, orange, horse-chestnut, pear, cherry, grape, pine and ash. Soak them in water for at least twenty-four hours. Plant them about 5 mm deep in pots containing a mixture half of sand and half of loam. Stand the pots in a warm place and keep the mixture moist. Date stones will grow if they are kept very warm and moist. They should be planted about 2 cm deep.

The growth of a twig

Consider the seasonal changes in the growth of a twig. A horse-chestnut twig is a good example. During spring, the *terminal bud* at the tip of the twig opens and the flowers and leaves contained in it open out and grow. In the summer, the flowers fade and fruits are formed. In autumn, the leaves wither and fall, and *leaf scars* are left behind. The *girdle scars* on the twig indicate the positions of buds of previous years. The distance between two successive girdle scars represents one year's growth. The small buds which occur beneath the terminal bud are called *lateral buds.* They grow and become side shoots.

A horse-chestnut twig

Examine and make a labelled drawing of a horse-chestnut twig. Late winter is the best time of the year for this. Notice the terminal bud, lateral buds, leaf scars, vein endings, girdle scars and lenticels (breathing pores). Break open one of the buds and look at the tightly-packed leaves inside it.

Twigs of common trees

Examine and make labelled drawings of twigs of some common trees, such as the ash, beech, etc.

PLANT GROWTH

AUTUMN AND WINTER
Breathing and transpiration at a minimum. Leaves fallen.

SPRING AND SUMMER
Sap rising. Breathing and transpiration at a maximum. Growth of leaves and flowers.

LEAF FALL

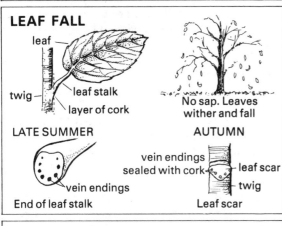

leaf
twig
leaf stalk
layer of cork

LATE SUMMER

No sap. Leaves wither and fall

AUTUMN

vein endings
sealed with cork
leaf scar
twig

vein endings

End of leaf stalk

Leaf scar

KNOTS IN WOOD

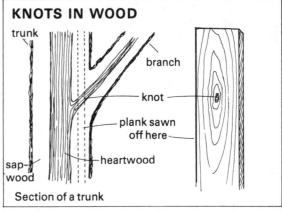

trunk
branch
knot
plank sawn off here
heartwood
sapwood

Section of a trunk

ANNUAL RINGS

LABURNUM

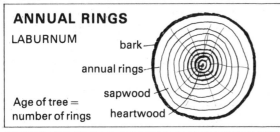

bark
annual rings
sapwood
heartwood

Age of tree = number of rings

LATERAL TWIG

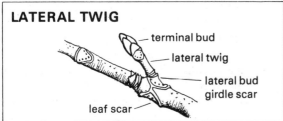

terminal bud
lateral twig
lateral bud
girdle scar
leaf scar

HORSE-CHESTNUT BUD

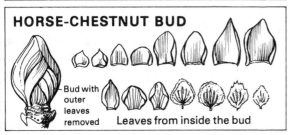

Bud with outer leaves removed

Leaves from inside the bud

HORSE-CHESTNUT TWIG

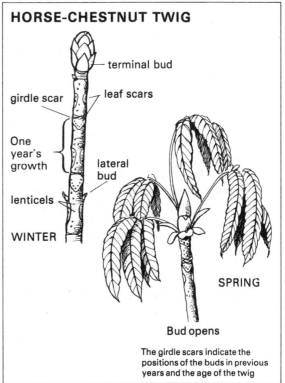

terminal bud
girdle scar
leaf scars
One year's growth
lateral bud
lenticels
WINTER

SPRING

Bud opens

The girdle scars indicate the positions of the buds in previous years and the age of the twig

4 Plant Growth

Plant growth

The growth of roots and shoots occurs at their tips. The growth of a root, in length, takes place in the region just behind the tip. The tip of a shoot is more sensitive to light than are its other parts. Shoots grow towards sunlight. This growth movement is called *phototropism*. Roots grow downwards. This movement is called *geotropism*. Roots also grow towards water. This movement is called *hydrotropism*. "Photo" means "light", "geo" means "earth" and "hydro" means "water".

Root growth

Use a hair dipped in Indian ink to make equally spaced marks, about 3 mm apart, along the root of a broad bean seedling. Seedlings can be grown on wet flannel in a jam-jar. After a few days of growth, you will notice that the spaces between the marks have increased in size only in the region behind the root tip.

Geotropism

Germinate mustard or radish seeds on a layer of moist cotton wool. Support the seeds on a wooden board and cover them with a small sheet of glass, about 10 cm × 10 cm. Use rubber bands to hold the glass in position. Allow the board to stand for a time, resting on its edge. Then when the roots are about 1 cm long, tilt the board through 90°. After a day or so, the roots will be found to be growing downwards again.

Hydrotropism

Place a porous pot containing water in the middle of a box filled with sawdust. Sawdust is cleaner than soil. Plant some soaked peas in the sawdust. Examine the peas after a few days. Their roots will have grown sideways towards the moist region around the pot.

Phototropism

Cover a pot containing a growing plant with a cardboard box that has a small hole in one of its sides. Remove the box after a few days. The plant will have grown towards the hole in the box.

Plants grown in pots indoors should be turned around periodically. Do you know why?

Shoot tips are sensitive to light

Grow several oat or wheat seedlings in a shallow dish. Cover some of the shoot tips with foil caps. Place a cardboard box with a small hole in one side over the seedlings. Remove the box after a few days. The shoots without foil caps will be bent over; they will have grown towards the hole in the box. The shoots with foil caps will still be upright.

Growing plants indoors

Here are some suggestions for growing plants indoors.
1. Cut a carrot in half. Scoop out the centre of the portion which has leaves. Fill the hole with water and, with the leaves underneath, suspend the carrot with cotton. The leaves grow and bend upwards.
2. Fill a dish with moist fibre. Place bulbs and corms in the fibre so that their shoots are just showing. Stand the dish in a warm dark place. Bring it out into the light when the shoots have begun to grow.
3. Make small holes in the skin of a marrow. Place small seeds in these holes. The seeds germinate and grow. Their roots obtain water from the marrow.
4. Place pieces of potato with eyes in a shallow dish containing water. When they sprout, transfer them to a pot containing soil.
5. Place carrot, parsnip and turnip tops in a shallow dish containing water. Cover the tops with clean pebbles. The pebbles hold the growing plants in place.
6. Cut a dandelion root into 2 cm lengths and plant them about 5 mm deep in a dish of soil. Keep the soil warm and moist. Each piece of root grows into a dandelion plant.
7. During early spring, put twigs of ash, horse-chestnut, etc., in a jar of water. Stand the jar in a warm place. The buds open to display their flowers and foliage.

PLANT GROWTH

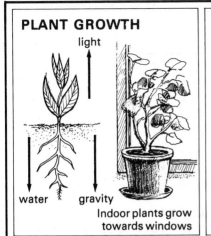

light

water • gravity

Indoor plants grow towards windows

ROOT GROWTH

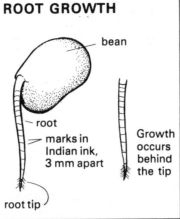

bean

root

marks in Indian ink, 3 mm apart

Growth occurs behind the tip

root tip

GEOTROPISM

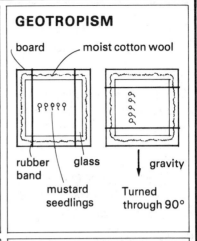

board • moist cotton wool

rubber band • glass

mustard seedlings

gravity

Turned through 90°

HYDROTROPISM

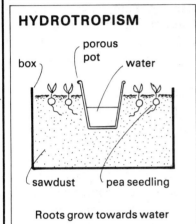

porous pot • water

box

sawdust • pea seedling

Roots grow towards water

PHOTOTROPISM

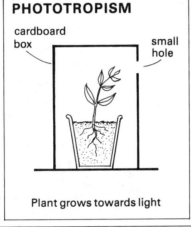

cardboard box

small hole

Plant grows towards light

SHOOT TIPS ARE LIGHT-SENSITIVE

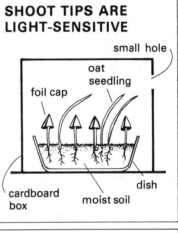

small hole

oat seedling

foil cap

cardboard box

moist soil

dish

GROWING PLANTS INDOORS

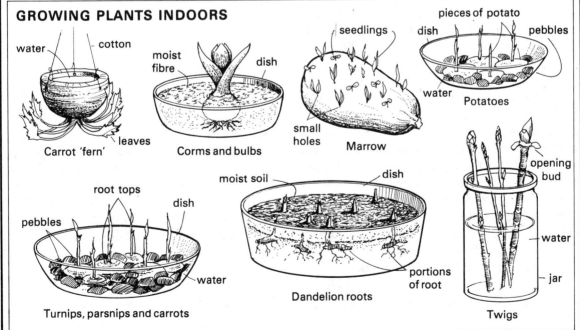

water • cotton

leaves

Carrot 'fern'

moist fibre • dish

Corms and bulbs

seedlings

small holes • Marrow

pieces of potato

dish • pebbles

water

Potatoes

pebbles

root tops • dish

water

Turnips, parsnips and carrots

moist soil • dish

portions of root

Dandelion roots

opening bud

water

jar

Twigs

5 | *Measuring Volume*

Volume

Volume, or *capacity,* means "space occupied". It is measured in *cubic metres* (m^3) and *cubic centimetres* (cm^3). Liquids, such as petrol and water, may also be measured in *litres* (ℓ) and *millilitres* ($m\ell$).

$$1\,m^3 = 1\,000\,000\,cm^3$$
$$1\ell = 1000\,m\ell = 1000\,cm^3$$
$$1\,m\ell = 1\,cm^3$$

Kitchen measures

In the kitchen, liquids are usually measured in litres and millilitres. 1 litre of water has a mass of 1 kilogram (kg). 1 millilitre of water has a mass of 1 gram (g).

Measuring volumes

Support a *burette* in a stand. Fill the burette with water. Then open the tap so that water runs out into a beaker. Close the tap when the water level has fallen to the zero mark near the top of the burette. Your eye, the zero mark and the bottom of the *meniscus* must be in line. Now empty the beaker and then open the tap again. Close it when the meniscus has fallen to the 17 cm^3 mark. You have delivered 17 cm^3 of water into the beaker. Open the tap once more. Close it when the meniscus has fallen to the 28 cm^3 mark. You have now delivered an extra 11 cm^3 of water into the beaker.

Pipettes are used for providing exact volumes of liquids—10 cm^3, 25 cm^3, 50 cm^3, etc. Suck water from a beaker into a 25 cm^3 pipette until its level is higher than the mark on the stem. Place your finger over the top end of the pipette stem. Then move your finger and allow air to enter the stem until the water has fallen to the level of the mark. The pipette now contains 25 cm^3 of water. Deliver this into a beaker by removing your finger from the top of the pipette stem. *Do not use a pipette for measuring poisonous liquids.*

Use a *measuring jar* to obtain different volumes of water. Simply pour water into the jar until the bottom of its meniscus has reached the appropriate graduation mark on the side of the jar.

Find the volume of a rectangular block of wood by measuring its length, breadth and height with a ruler.

$$Volume = length \times breadth \times height$$

The volume of a pebble

The volumes of irregular solids are found by using a measuring jar. Pour enough water to cover a pebble into a measuring jar and note its volume. Immerse a pebble in the water. Note the new volume. Subtraction gives the volume of the pebble.

Volume of water + pebble	$= x\,cm^3$
Volume of water	$= y\,cm^3$
$\therefore$ Volume of pebble	$= (x-y)\,cm^3$

The volume of a cork

The volumes of irregular solids which float are found by using a *sinker*. Pour some water into a measuring jar and note its volume. Immerse a cork, attached by cotton to a pebble of known volume, in this water. The pebble is the sinker. Note the new volume.

Volume of water + pebble + cork	$= x\,cm^3$
Volume of water	$= y\,cm^3$
Volume of pebble	$= z\,cm^3$
Volume of water + pebble	$= (y+z)\,cm^3$
$\therefore$ Volume of cork	$= x-(y+z)\,cm^3$

A volume of sugar

The volumes of substances which dissolve in water are found by using a liquid in which they do not dissolve. Pour some methylated spirit into a measuring jar and note its volume. Immerse a quantity of sugar in the spirit and note the new volume. Sugar does not dissolve in methylated spirit.

Volume of spirit + sugar	$= x\,cm^3$
Volume of spirit	$= y\,cm^3$
$\therefore$ Volume of sugar	$= (x-y)\,cm^3$

SOME IMPORTANT UNITS OF LENGTH AND VOLUME

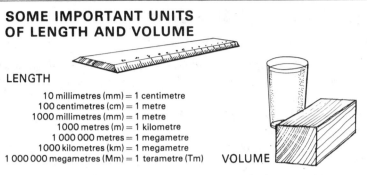

LENGTH

10 millimetres (mm) = 1 centimetre
100 centimetres (cm) = 1 metre
1000 millimetres (mm) = 1 metre
1000 metres (m) = 1 kilometre
1 000 000 metres = 1 megametre
1000 kilometres (km) = 1 megametre
1 000 000 megametres (Mm) = 1 terametre (Tm)

VOLUME

1000 millilitres (ml) = 1 litre (l)
1 millilitre = 1 cubic centimetre
1 000 000 cubic centimetres (cm³) = 1 cubic metre (m³)
1000 cubic centimetres = 1 cubic decimetre
1000 cubic decimetres (dm³) = 1 cubic metre
1000 millilitres = 1 cubic decimetre
1 litre = 1 cubic decimetre

VOLUME

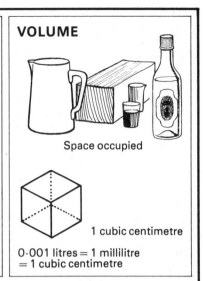

Space occupied

1 cubic centimetre

0·001 litres = 1 millilitre
= 1 cubic centimetre

KITCHEN MEASURE

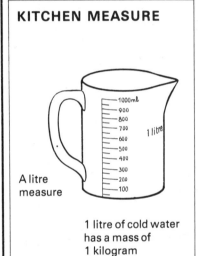

A litre measure

1 litre of cold water has a mass of 1 kilogram

MEASURING VOLUMES

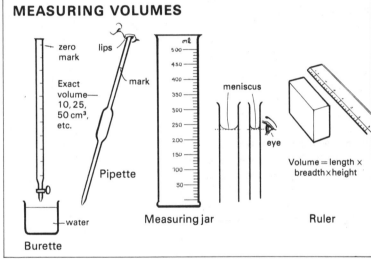

zero mark

lips

Exact volume— 10, 25, 50 cm³, etc.

mark

Pipette

water

Burette

ml
500
450
400
350
300
250
200
150
100
50

meniscus

eye

Measuring jar

Volume = length × breadth × height

Ruler

THE VOLUME OF A PEBBLE

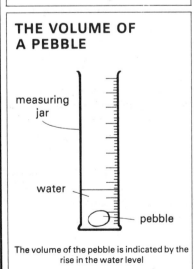

measuring jar

water

pebble

The volume of the pebble is indicated by the rise in the water level

THE VOLUME OF A CORK

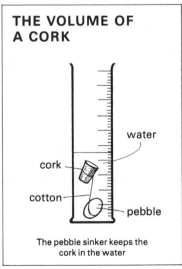

water

cork

cotton

pebble

The pebble sinker keeps the cork in the water

A VOLUME OF SUGAR

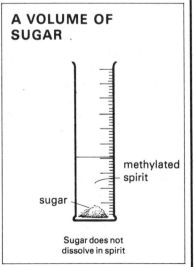

methylated spirit

sugar

Sugar does not dissolve in spirit

Mass

The *mass* of an object is the quantity of material it contains. A volume of iron has a much greater mass than an equal volume of wood, and a mass of iron occupies a much smaller space than an equal mass of wood. Mass is measured in *kilograms* (kg) and *grams* (g). 1 kg = 1000 g.

Gravity, force and weight

The pull, or *force,* of the Earth's *gravity* on an object gives it *weight*. Forces are measured in *newtons* (N). The gravitational force on a mass of 1 kilogram, which is 1 kilogram force (1 kgf), is about 10 newtons. Since all objects on the Earth are subjected to its gravitational force, weighing is a way of comparing masses. Balances are usually marked in grams, which are units of mass, but balances really measure weight, or gravitational force.

$$10 \text{ N} \simeq 1 \text{ kgf}$$
$$1 \text{ kgf} = 1000 \text{ gf}$$
$$1 \text{ N} \simeq 100 \text{ gf}$$

Finding the mass of water

Weigh a litre measure on a spring balance. Pour water into the measure until it is full and then reweigh it. Subtraction gives the mass of one litre of water. 1 litre of cold water has a mass of 1 kg.

$$\text{Mass of measure + water} = x \text{ g}$$
$$\text{Mass of measure} = y \text{ g}$$
$$\text{Mass of water} = (x - y) \text{ g}$$

Making a balance

Cut notches at distances of 1 cm from the ends of a thin wooden rod of uniform thickness and about 50 cm long. Use strong thread to suspend pans, made of equal-sized metal lids, from these notches. Cut a notch in the middle of the rod. Attach a loop of string to this notch. If the rod does not balance, that is, if the rod does not take up a horizontal position when the loop of string is held in the fingers, then one side of the rod is heavier than the other. This can be adjusted by attaching a small movable coil of wire to the lighter side of the rod. Move this wire until an exact balance is obtained.

Make plasticine masses of 0.5 g, 1 g, 2 g, 5 g, 10 g, 20 g and 50 g. Compare them with standard metal masses. Weigh various objects.

A model spring balance

Fix a steel spring, with a pan and a pointer attached, to a wooden lath held in a clamp. Attach a paper scale to the lath. Calibrate the scale in grams by placing standard metal masses, one at a time, in the pan. Do not over-strain the spring.

The common balance

One of the illustrations shows a *common balance*. At one time it was widely used in laboratories for weighing small quantities of substances with great accuracy. Unfortunately, though this instrument is very accurate, it is very delicate and is easily damaged; also, it needs to be levelled and adjusted correctly before it can be used. In recent years, electronic balances have become popular, and it is likely that you will make accurate measurements of mass and weight with one of these.

An electronic balance

An electronic balance is operated electrically. It has certain advantages: it is completely enclosed, it is not easily damaged and it is not heavy, so it can be easily carried about; it has no moving parts, so there is very little to go wrong; it is very accurate and can measure very small masses and weights; the measurements are displayed on a dial in large and bright figures that can be easily read at a distance.

Perhaps your teacher will show you how to operate an electronic balance. Use it for measuring the masses of some small objects.

Making a letter balance

Use wire to fix a rubber band to the top of a piece of thick cardboard. Attach a wire with a hook to the bottom of the rubber band. Calibrate the scale by suspending masses of 10 g, 20 g, 30 g, etc., in turn, from the hook. Weigh letters and small objects.

BALANCES

Earth's pull— gravity

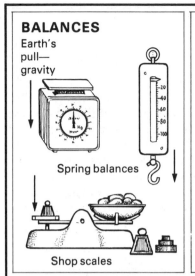

Spring balances

Shop scales

SOME UNITS OF MASS AND WEIGHT

MASS

1000 milligrams (mg) = 1 gram
1000 grams (g) = 1 kilogram
1 000 000 grams = 1 megagram (Mg)
1000 kilograms (kg) = 1 megagram
1000 kilograms = 1 tonne (t)

WEIGHT

1000 grams force (gf) = 1 kilogram force (kgf)
10 newtons (N) $\simeq$ 1 kilogram force
1 newton $\simeq$ 100 grams force
1000 kilograms force = 1 *tonne*

THE MASS OF WATER

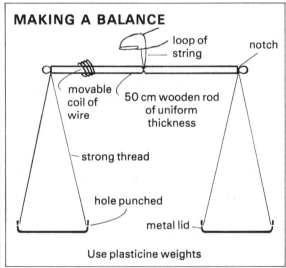

Weigh litre measure empty

Pour water into the measure and reweigh it

MAKING A BALANCE

loop of string
notch
movable coil of wire
50 cm wooden rod of uniform thickness
strong thread
hole punched
metal lid

Use plasticine weights

MODEL SPRING BALANCE

clamp
drawing-pin
hole
wooden lath
spring
paper scale
g
0
10
20
30
40
50
metal lid
wire pointer

COMMON BALANCE

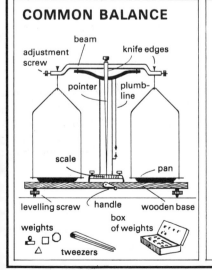

adjustment screw
beam
knife edges
pointer
plumb-line
scale
pan
levelling screw
handle
wooden base
weights
box of weights
tweezers

ELECTRONIC BALANCE

accurate, rapid in operation, clear readings, robust, enclosed, no moving parts, portable, foolproof

MAKING A LETTER BALANCE

wire
rubber band
cardboard 5 cm × 15 cm
g
− 10
− 20
− 30
− 40
− 50
scale
wire
hook
wire frame to hold letters

7 | *Density*

Density

Equal volumes of different materials do not have the same mass. Some materials are denser than others. Lead is much denser than water. *Density* is measured as mass per unit volume, that is, *kilograms per cubic metre* (kg/m³) or *grams per cubic centimetre* (g/cm³). 1 cm³ of water has a mass of 1 g, 1 cm³ of lead has a mass of 11.3 g, and 1 cm³ of alcohol has a mass of only 0.82 g, and their densities are: water, 1 g/cm³; lead, 11.3 g/cm³; alcohol, 0.82 g/cm³.

The density of water

Show that 1 cm³ of water has a mass of 1 g, that is, the density of water is 1 g/cm³. Weigh a can on an electronic balance. Reweigh it after pouring in 100 cm³ of water from a measuring jar. Why is it better to weigh 100 cm³ of water and then divide by 100 than to weigh just 1 cm³?

$$
\begin{aligned}
\text{Mass of water} + \text{can} &= x\,\text{g} \\
\text{Mass of can} &= y\,\text{g} \\
\therefore \text{Mass of water} &= (x-y)\,\text{g} \\
\therefore \text{Mass of 1 cm}^3\text{ water} &= \frac{(x-y)}{100}\,\text{g}
\end{aligned}
$$

Finding densities of solids

Find the volumes of rectangular solids of iron, clay and wood using a ruler marked in centimetres. Weigh these solids on an electronic balance. Calculate their densities.

$$\text{Density} = \frac{\text{mass}}{\text{volume}}\ \text{g/cm}^3$$

Find the volumes of irregular solids, such as pebbles, granite chippings and plasticine, with a measuring jar. Weigh them. Then calculate their densities.

Finding densities of liquids

Weigh an empty dish. Reweigh the dish when it contains methylated spirit delivered from a 50 mℓ pipette. Subtraction gives the mass of the spirit. Now determine the densities of sea-water and paraffin.

$$
\begin{aligned}
\text{Mass of dish} + \text{liquid} &= x\,\text{g} \\
\text{Mass of dish} &= y\,\text{g} \\
\text{Mass of liquid} &= (x-y)\,\text{g} \\
\text{Density of liquid} &= \frac{(x-y)}{50}\,\text{g/cm}^3
\end{aligned}
$$

Relative density

Relative density (which is sometimes called *specific gravity*) is the number of times a volume of material is denser than an equal volume of water. 1 cm³ of water has a mass of 1 g. Therefore, water has a relative density of 1. 1 cm³ of lead has a mass of 11.3 g. Lead is 11.3 times denser than water. Therefore, lead has a relative density of 11.3. Alcohol has a relative density of 0.82.

$$\text{Relative density} = \frac{\text{mass of material}}{\text{mass of an equal volume of water}}$$

Hydrometers

Hydrometers are instruments used for finding the relative densities of liquids. The level of a hydrometer in a liquid indicates its density. Hydrometers are used for finding the relative density of the sulphuric acid in accumulators. They are used for determining the relative densities of wines, spirits, beer and milk. This practice deters unscrupulous people from adding water to milk and alcoholic beverages.

Making a hydrometer

Make a hydrometer by placing a paper scale inside a test-tube. Fill the bottom of the test-tube with lead shot. Fix the shot in place with molten paraffin wax. The weight of this shot will hold the test-tube upright in a liquid. Insert a cork in the test-tube. Float the hydrometer in water. Note on the paper scale the level to which it falls. Now float the hydrometer in spirit. You notice that the hydrometer falls to a lower level in the spirit. This hydrometer can be graduated by immersing it in different liquids of known relative densities.

Measure the relative densities of milk, paraffin and sea-water.

A wood hydrometer

One of the diagrams shows how a hydrometer can be made from a stick of wood.

DENSITY

MASS PER UNIT VOLUME

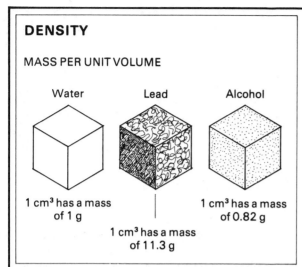

Water

1 cm³ has a mass of 1 g

Lead

1 cm³ has a mass of 11.3 g

Alcohol

1 cm³ has a mass of 0.82 g

RELATIVE DENSITY OF LEAD

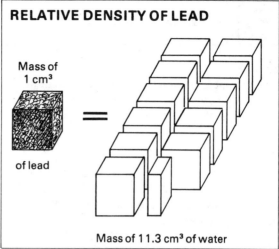

Mass of 1 cm³ of lead

=

Mass of 11.3 cm³ of water

SOME RELATIVE DENSITIES

Gold	19.3
Mercury	13.6
Copper	8.9
Iron	7.8
Cork	0.25
Oak	0.9
Silver	10.5
Lead	11.3
Alcohol	0.82
Water	1.0

COMMERCIAL HYDROMETER

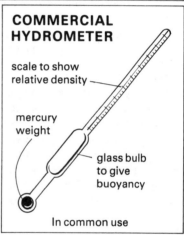

scale to show relative density

mercury weight

glass bulb to give buoyancy

In common use

MAKING A HYDROMETER

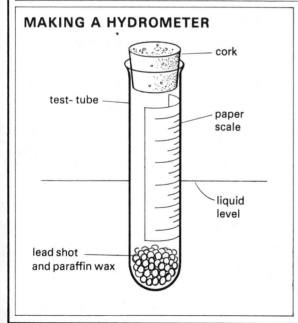

cork

test- tube

paper scale

liquid level

lead shot and paraffin wax

WOOD HYDROMETER

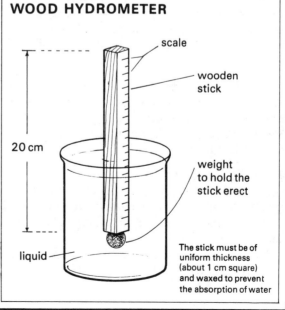

scale

wooden stick

20 cm

weight to hold the stick erect

liquid

The stick must be of uniform thickness (about 1 cm square) and waxed to prevent the absorption of water

8 | Buoyancy

Buoyancy
Objects appear to weigh less in liquids than in air because of *buoyancy,* or *upthrust.* This upthrust keeps ships afloat and swimmers on top of the water.

Objects weigh less in water
Hang a standard weight on the hook of a spring balance. Then suspend the weight in water. The weight shown by the balance is less than that marked on the weight.

Suspend two equal weights from the ends of a rod, pivoted at its mid-point, so that it is exactly balanced. Place one weight in water. The other end of the rod falls. The apparent loss in weight is caused by the buoyancy of the water.

Using buoyancy
An iron ship floats because only its hull is made of iron and, being filled with air, it weighs less than an equal volume of water. Submarines pump water in and out of their ballast tanks in order to descend and ascend. A balloon floats in the atmosphere because the gas which it contains is less dense than air. Swimmers find that salt water is more buoyant than fresh water.

Plimsoll lines
Plimsoll lines, which show the depths to which ships can safely sink in water when loaded, are named after Samuel Plimsoll, the man who devised them. Overloaded ships are liable to capsize.

A plasticine submarine
Make a small hollow cylinder of plasticine with both ends closed. Place it in water. It floats. Why? Now remove one end of the cylinder. Place it in water. Why does it sink?

Water displacement
The buoyancy of a liquid depends upon its density. Wood floats on water because its density is less than that of water. Oil floats on water. Iron sinks in water. Why?

An iron ship floats because the weight of its total volume, including the air inside it, equals the weight, or upthrust, of the water it displaces. Since the combined iron and air of which the ship is composed is not as heavy as the water which would occupy the same space as the ship, the volume of water displaced is less than the volume of the ship (see the diagram).

Finding the mass of a dish
Float an empty dish in a large bowl that is completely full of water. Catch the overflow on a tray. Pour this water into a measuring jar to find its volume, and, from this, its mass. 1 cm^3 of water has a mass of 1 g. The mass of the dish equals the mass of the water displaced.

Archimedes' Principle
When a solid is immersed in a liquid, the apparent loss in weight of the solid equals the weight of the liquid displaced. This is the *Principle of Archimedes.* Archimedes was a Greek who devoted his life to the study of science.

Finding relative density by Archimedes' Principle
Weigh a pebble, first in air, and then suspended in water. The apparent loss in weight of the pebble equals the weight of the water displaced. But the volume of the pebble equals the volume of the water displaced. Therefore,

$$\text{Relative density of pebble} = \frac{\text{weight of pebble}}{\text{loss in weight of pebble}}$$

Finding the relative density of an egg
Put an egg in a jar containing fresh water. It sinks because it is denser than the water. Put the egg in a strong solution of salt water. It floats because it is less dense than salt water. Then put the egg in fresh water again. Add salt and stir until the egg floats just below the surface. Its density is now the same as that of the salt water. Use a hydrometer to find the relative density of the salt water and so of the egg.

Bad eggs float
A test for eggs is to place them in water. Bad eggs usually float because they are less dense than water.

OBJECTS WEIGH LESS IN WATER

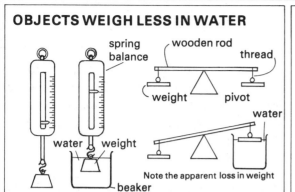

spring balance

wooden rod

thread

weight — pivot

water

water weight

beaker

Note the apparent loss in weight

FINDING RELATIVE DENSITY

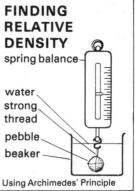

spring balance

water
strong thread
pebble
beaker

Using Archimedes' Principle

ARCHIMEDES

287-212 B.C.

He founded the science of mechanics

USING BUOYANCY

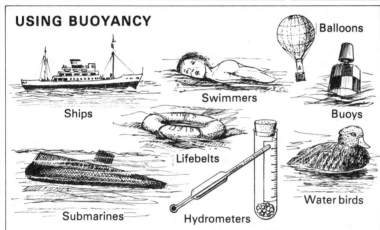

Ships

Swimmers

Balloons

Buoys

Lifebelts

Submarines

Hydrometers

Water birds

PLIMSOLL LINES

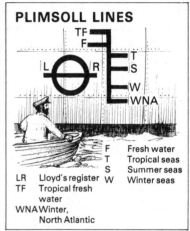

TF
F
L   R   T
S
W
WNA

LR — Lloyd's register
TF — Tropical fresh water
WNA — Winter, North Atlantic

F — Fresh water
T — Tropical seas
S — Summer seas
W — Winter seas

WHY IRON SHIPS FLOAT

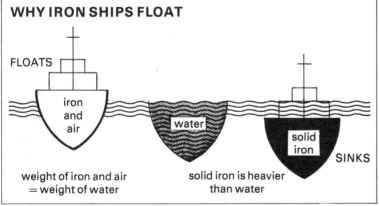

FLOATS

iron and air

water

solid iron

SINKS

weight of iron and air = weight of water

solid iron is heavier than water

PLASTICINE SUBMARINE

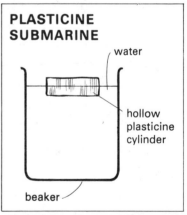

water

hollow plasticine cylinder

beaker

THE RELATIVE DENSITY OF AN EGG

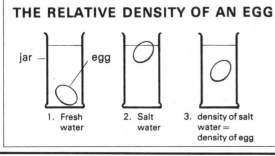

jar — egg

1. Fresh water

2. Salt water

3. density of salt water = density of egg

BAD EGGS FLOAT

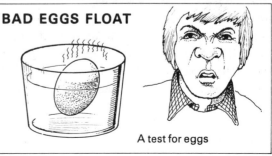

A test for eggs

Water pressure

The deeper a diver goes, the larger is the amount of water above him, and so the greater is the pressure on his body.

Divers wear thick, reinforced suits so that they are not crushed by water pressures. The *bathysphere* in which Dr. Beebe explored the ocean depths in 1934 had thick metal walls to withstand enormous water pressures. Professor Piccard's *bathyscaphe,* in which he descended two miles in the Mediterranean Sea, was of a similar construction. Fish which live at great ocean depths are rarely seen on the surface; their bodies have been adapted by nature to withstand great pressures. The walls of reservoirs and dams are thickest at their bases, for it is there that the water pressures are greatest.

Pressure and depth

Imagine a large cylinder which contains 8000 kg of water. The *force,* or *thrust,* pressing on the bottom of the cylinder is 8000 kgf.

But pressure is measured as force per unit area, that is, in *newtons per square metre.* Therefore, in this case, if the cylinder has a base area of 2 m^2, the pressure of the water is 40 000 N/m^2.

$$\text{Pressure} = \frac{\text{force}}{\text{area}} = \frac{8000\,\text{kgf}}{2\text{m}^2}$$

$$= 4000\,\text{kgf/m}^2 \simeq 40\,000\,\text{N/m}^2$$

Force (or thrust) = pressure × area

Water pressure increases with depth

Put a small quantity of mercury into the bend of a U-tube which has a long arm and a short arm. Lower the short arm of the tube into a jar of water. The mercury in the long arm rises. The pressure on the mercury increases with the depth of the water.

Hold a piece of thick card against the end of a wide glass tube and lower it into a large beaker of water. The card is held in position by water pressure. Pour water slowly into the tube. When the height of the water in the tube is just greater than that in the beaker, the card falls away. Why?

Water pressure is equal in all directions

Attach a very short rubber tube to the shorter arm of the U-tube apparatus. Lower the tube into a large beaker of water. Point this rubber tube in different directions. The mercury does not move. This shows that the pressure is the same in all directions.

A self-filling bucket

When a self-filling bucket is pushed into water, a valve is forced open. When the bucket is carried away, the pressure of the water in it keeps the valve closed.

The hydraulic press

The *hydraulic press* is used for moving heavy loads. An aeroplane undercarriage is lowered and raised by hydraulics, though the control device may be electrically operated.

The hydraulic principle

10 N of effort pressing on 1 cm^2 has the same pressure as a 1000 N load pressing on 100 cm^2. But when the water falls by 1 cm, the load rises by only $\frac{1}{100}$ cm. *What is gained in force is lost in distance.*

A model hydraulic press

You can show how the hydraulic press operates by using the apparatus shown in the diagram. A few centimetres of water in the long arm of the tube is sufficient to move the 0.5 kgf weight. This weight moves when the pressures in the two arms of the tube are equal.

WATER PRESSURE

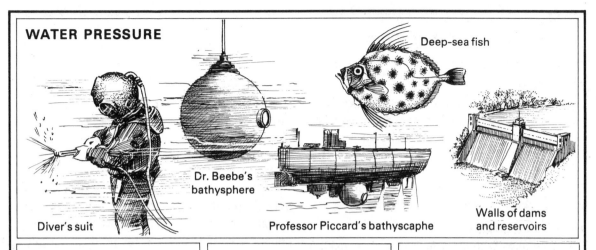

Diver's suit

Dr. Beebe's bathysphere

Professor Piccard's bathyscaphe

Deep-sea fish

Walls of dams and reservoirs

PRESSURE AND DEPTH

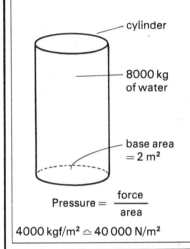

cylinder

8000 kg of water

base area = 2 m²

$$\text{Pressure} = \frac{\text{force}}{\text{area}}$$

4000 kgf/m² ≏ 40 000 N/m²

WATER PRESSURE INCREASES WITH DEPTH

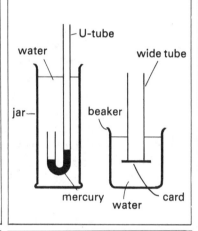

U-tube

water

wide tube

jar

beaker

mercury

water

card

WATER PRESSURE IS EQUAL IN ALL DIRECTIONS

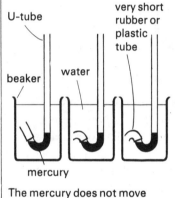

U-tube

very short rubber or plastic tube

beaker

water

mercury

The mercury does not move

SELF-FILLING BUCKET

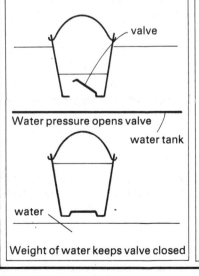

valve

Water pressure opens valve

water tank

water

Weight of water keeps valve closed

HYDRAULIC PRINCIPLE

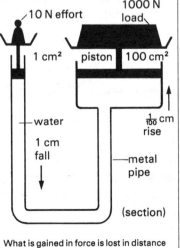

10 N effort

1000 N load

1 cm²

piston 100 cm²

water

1 cm fall

$\frac{1}{100}$ cm rise

metal pipe

(section)

What is gained in force is lost in distance

MODEL PRESS

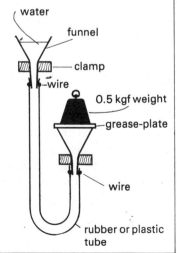

water

funnel

clamp

wire

0.5 kgf weight

grease-plate

wire

rubber or plastic tube

10 | *Air Pressure*

Air is elastic

Plug the open end of a bicycle pump with plasticine. Push the handle forward quickly. The air in the barrel is compressed. The plasticine is expelled and the compressed air is released with a *pop*.

Air, then, after being *compressed,* returns to its original volume. It is *elastic.*

Air presses in all directions

Tyres are filled with air, which acts as a cushion or a spring. When a football is kicked, it loses its shape, but only temporarily, for the compressed air in the ball forces the case back into its normal shape. A football bounces because the air which it contains acts as a spring.

Balloons, footballs, tennis balls, etc., which are filled with air under pressure, keep their shapes. Air presses equally in all directions. It is for this reason that soap bubbles are spherical in shape.

The atmospheric ocean

An enormous pressure is exerted by the water at the bottom of an ocean. In the same way the total mass of the atmosphere, which is an ocean of air over 300 kilometres in depth, is pressing on the Earth's surface.

Air has weight

Place a balloon on the pan of an electronic balance. Make a note of its weight. Then inflate the balloon and put it back on the pan. Has its weight increased?

The enormous pressure of the atmosphere

Heat a thin-walled metal can containing a small quantity of water. A petrol can would do for this, but make sure that all the petrol has been removed. Close the can when the water is boiling. The can must be air-tight. Allow the can to cool. The steam inside the can condenses back into water and the walls of the can are crushed inwards by the force exerted by the atmosphere.

Showing air pressure

Completely fill a tall jar with water and cover its mouth with a glass cover or a piece of cardboard. Slowly invert the jar. The water remains in the jar and the cover is held in place by atmospheric pressure.

Fill a tall jar with water and place a glass cover over its mouth. Invert the jar and stand it in a trough containing water. Remove the glass cover under water. The water in the jar does not fall. It is held in place by the force of the atmosphere pressing downwards on the water in the trough.

Burn a small quantity of paper in a wide-necked bottle. When the paper has finished burning, insert a hard-boiled shelled egg into the neck of the bottle. What happens? The burning paper causes the air in the bottle to become hot and expand. Some of it escapes from the bottle. When the remaining air cools, it contracts and the egg is forced into the bottle by atmospheric pressure.

Dip a glass tube into water and cover the top end with your finger. Lift the tube. The water in it does not fall out. Why?

The Magdeburg hemispheres

Two copper hemispheres were used by Otto von Guericke, a German scientist, to show the enormous pressure exerted by the atmosphere. His famous experiment was demonstrated to the German Emperor at Magdeburg, in Saxony, in 1651. A strong copper sphere was made in two halves. They were placed together and the air inside them was pumped out. Sixteen horses, eight on each side, were required to pull the hemispheres apart.

A vacuum

A space which contains no air or any other gases is called a *vacuum.*

AIR IS ELASTIC

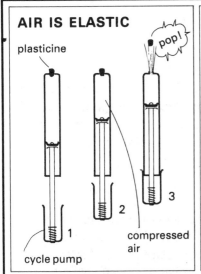

plasticine

pop!

cycle pump

1   2   3

compressed air

AIR PRESSES IN ALL DIRECTIONS

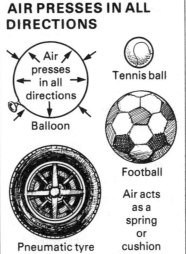

Air presses in all directions

Balloon

Tennis ball

Football

Pneumatic tyre

Air acts as a spring or cushion

OTTO VON GUERICKE

1602-1686

THE ENORMOUS PRESSURE OF THE ATMOSPHERE

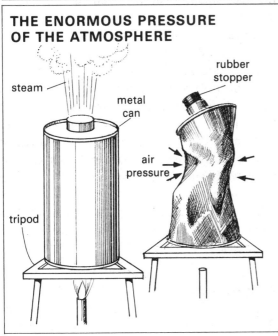

steam

metal can

rubber stopper

air pressure

tripod

SHOWING AIR PRESSURE

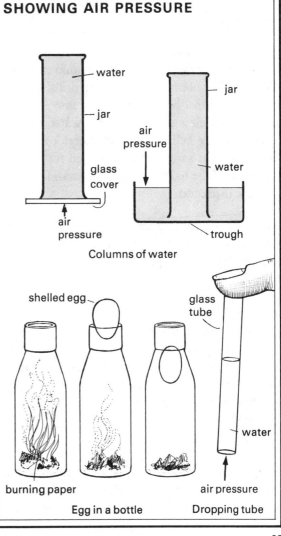

water

jar

glass cover

air pressure

jar

air pressure

water

trough

Columns of water

shelled egg

glass tube

burning paper

Egg in a bottle

water

air pressure

Dropping tube

MAGDEBURG HEMISPHERES

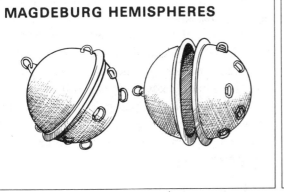

Using air pressure

Many useful devices work by air pressure. Shop window suckers, dropping tubes, syringes, pipettes, pumps, siphons, drinking straws, animal drinking fountains and preserving jars are operated by air pressure.

When you suck at a drinking straw, the air in it is drawn into your mouth. The drink, under atmospheric pressure, rushes up the straw to take the place of this air and enters your mouth. Lids are placed on jars when their contents are hot. The air and vapours trapped under the tops of the jars contract on cooling and the atmosphere exerts a pressure on their lids. These lids are effective seals and the contents of the jars remain in a good condition.

Air under pressure is used in vacuum cleaners, diving bells, submarines, pressure drills, certain types of lifts, liquid sprayers and brakes. Compressed air is used to keep the water at a low level in diving bells so that engineers can work inside them in safety. It is also used for forcing water out of the ballast tanks of submarines when they are to ascend.

An automatic flushing tank

One of the diagrams shows a section of an automatic flushing tank of the kind used in public lavatories.

A model flushing tank

Use a bell-jar and a U-tube to make a model flushing tank in the way shown opposite.

Fill the jar with water dripping slowly from a tap. What happens when the water in the jar reaches the top of the U-tube (which is acting as a siphon)?

A toy sprinkler

Use a hammer and a nail to punch many small holes in the bottom of a metal can. Punch a single hole in the centre of the lid. Fill the can with water and replace the lid. Cover the hole in the top of the lid with your finger; the sprinkler ceases to operate. Remove your finger; the sprinkler operates again. This sprinkler can be used for watering indoor plants.

Pumps

In the *lift pump* shown opposite, the *piston* is raised and the weight of the water above it keeps the *valve* A closed. Air pressure forces water up the inlet pipe through the open valve B into the *barrel*. Water leaves the outlet pipe. On a downward stroke, the water below the piston closes the valve B and opens the valve A so that water fills the part of the barrel above the piston.

In the force pump shown opposite, an upward stroke decreases the air pressure in the barrel and the valve C closes. Valve D opens and water, under atmospheric pressure, enters the barrel. On a downward stroke, the water in the barrel closes the valve D and opens the valve C. Water leaves the outlet pipe.

Operate a syringe and glass models of these pumps if they are available. Watch the action of the valves.

A bicycle pump

The valve of a bicycle pump is a pliable washer. When the handle is pushed forward, the air in the barrel is compressed, the washer is flattened against the sides of the barrel, and the air is pushed out of the pump. When the handle is pulled out, the washer bends, under atmospheric pressure, and air fills the barrel. This pump is used for inflating bicycle tyres, footballs, etc.

Operate a bicycle pump if one is available. Press your finger against the end of the connector. Can you feel the force of the compressed air?

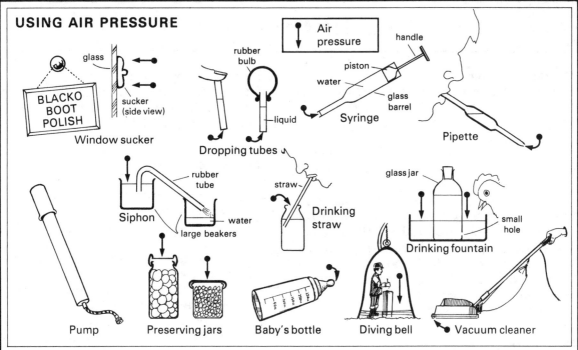

USING AIR PRESSURE

Air pressure

Window sucker
glass — sucker (side view)
BLACKO BOOT POLISH

Dropping tubes
rubber bulb — liquid

Syringe
handle — piston — water — glass barrel

Pipette

Siphon
rubber tube — water — large beakers

Drinking straw
straw

Drinking fountain
glass jar — small hole

Pump

Preserving jars

Baby's bottle

Diving bell

Vacuum cleaner

AUTOMATIC FLUSHING TANK

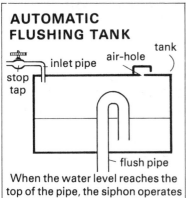

tank — air-hole — inlet pipe — stop tap — flush pipe

When the water level reaches the top of the pipe, the siphon operates

MODEL FLUSHING TANK

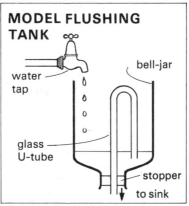

water tap — bell-jar — glass U-tube — stopper — to sink

TOY SPRINKLER

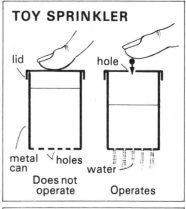

lid — hole — metal can — holes — water

Does not operate   Operates

LIFT PUMP

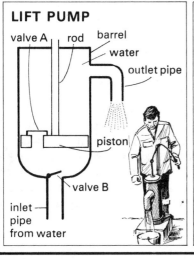

valve A — rod — barrel — water — outlet pipe — piston — valve B — inlet pipe from water

FORCE PUMP

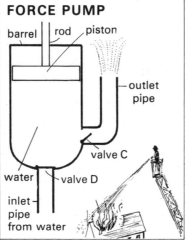

barrel — rod — piston — outlet pipe — valve C — water — valve D — inlet pipe from water

BICYCLE PUMP (section)

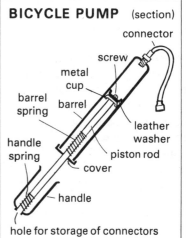

connector — screw — metal cup — barrel spring — barrel — handle spring — leather washer — piston rod — cover — handle — hole for storage of connectors

Measuring pressure

Pressure, which was mentioned on page 24, is force per unit area, and it can be measured in newtons per square metre (N/m^2). In fact, the proper unit of pressure, whether it be air pressure, water pressure or any other kind of pressure, is the *pascal* (Pa). 1 pascal = 1 newton per square metre, or $1\,Pa = 1\,N/m^2$.

The atmosphere exerts a pressure, and each square centimetre of a surface exposed to the atmosphere has a force of about 10 N (about 1 kgf) acting upon it. A piece of paper, 2 cm square, has a force of 40 N (about 4 kgf) acting on it; a force of 10 N acts on each square centimetre of the 4 cm^2 of paper. The normal pressure of the atmosphere is about 100 000 Pa, which is 100 000 N/m^2. Another unit, called the *bar,* is also used for measuring atmospheric pressure. 1 bar = 100 000 Pa.

The pressure of the atmosphere

The Italian scientist Galileo discovered that air pressure will support a column of water about 10 m high. A long tube with a cross-sectional area of 1 cm^2 and over 10 m high, and closed at one end, standing in a tank of water, would contain about 1.03 kg of water. It is for this reason that a lift pump does not raise water from a depth of more than about 10 m. Lift pumps are used for raising water from wells and tanks in the ground.

Torricelli, who was one of Galileo's students, found that the atmosphere would support a column of mercury about 76 cm high. Mercury is much denser than water.

Making a mercury barometer

Fill a glass tube, 80 cm long and closed at one end, with mercury. Hold your thumb tightly over the open end of the tube and invert it into a small dish containing mercury. On removing your thumb, the mercury falls until it is about 76 cm high. *Now wash your hands, for mercury is poisonous.* When the air pressure falls, the mercury falls slightly, and when the air pressure rises, the mercury rises slightly. This instrument can, therefore, be used for measuring air pressure. The space at the top of the tube is perfectly empty. It is called a *Torricellian vacuum.* Tilt the glass tube. The space at the top becomes smaller but the mercury remains at a height of about 76 cm.

An aneroid barometer

Mercury barometers are not handy and so, for everyday purposes, *aneroid barometers* are used. Aneroid means "without liquid".

One of the diagrams shows a side section of an aneroid barometer. A pointer is attached by levers to a thin metal box from which most of the air has been removed. When the air pressure is high, the end of the box is forced inwards. When the air pressure is low, the box returns to its original position. The pointer shows these movements on a dial. Generally, low pressure means bad weather and high pressure means fine weather. A household barometer is usually marked *storm, rain, change, fair* and *fine*.

A barograph

A *barograph* is used for recording air pressures automatically. Perhaps you have seen one of these instruments in an optician's window. A pen attached to a barometer draws an ink graph on graph-lined paper attached to a drum which makes one complete rotation in a week.

Air pressure and height

Air is less dense at great *altitudes,* or heights. Therefore, barometers, marked in metres, can be used for measuring heights. Such instruments are called *altimeters.* A barometer falls 1 cm for every 120 m rise. The air pressure on the top of a hill 240 m high would be 74 cm of mercury if the pressure at sea level were 76 cm of mercury.

Measuring gas pressure

Use a rubber or plastic tube to connect a *manometer* to the gas supply. A manometer is a glass U-tube partly filled with water. Turn on the gas. The water level in one arm falls and the water level in the other arm rises. The difference in the levels gives the gas pressure in centimetres of water.

NORMAL ATMOSPHERIC PRESSURE

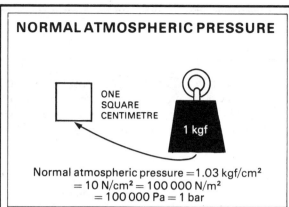

ONE SQUARE CENTIMETRE

1 kgf

Normal atmospheric pressure = 1.03 kgf/cm²
= 10 N/cm² = 100 000 N/m²
= 100 000 Pa = 1 bar

MEASURING AIR PRESSURE

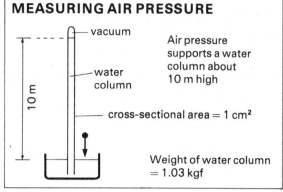

vacuum

water column

10 m

cross-sectional area = 1 cm²

Air pressure supports a water column about 10 m high

Weight of water column = 1.03 kgf

MERCURY BAROMETER

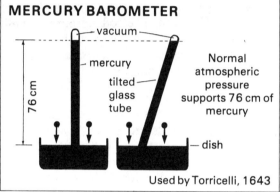

vacuum

mercury

tilted glass tube

76 cm

dish

Normal atmospheric pressure supports 76 cm of mercury

Used by Torricelli, 1643

BAROGRAPH

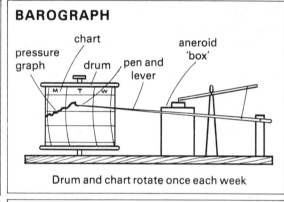

pressure graph

chart

drum

pen and lever

aneroid 'box'

Drum and chart rotate once each week

ANEROID BAROMETER
(section)

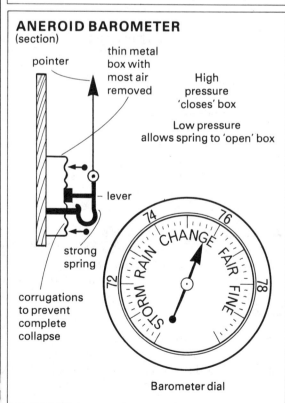

pointer

thin metal box with most air removed

High pressure 'closes' box

Low pressure allows spring to 'open' box

lever

strong spring

corrugations to prevent complete collapse

Barometer dial

AIR PRESSURE AND HEIGHT

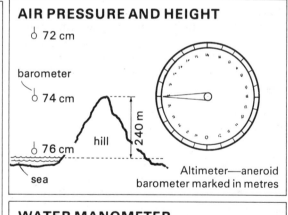

72 cm

barometer

74 cm

hill

76 cm

240 m

sea

Altimeter—aneroid barometer marked in metres

WATER MANOMETER

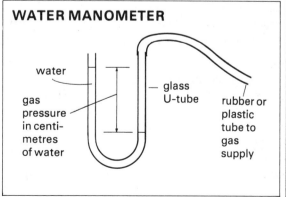

water

gas pressure in centimetres of water

glass U-tube

rubber or plastic tube to gas supply

13 | *Oxidation*

Oxidation and burning

When materials burn they combine with oxygen in the air to form *oxides*. This is called *oxidation*. Magnesium burns in oxygen to give *magnesium oxide*. Magnesium oxide contains oxygen and magnesium. Iron rusts in the presence of water and oxygen to give a brown rust called *iron oxide*. Rusting can be regarded as very slow burning. Oxygen oxidizes some of the carbon in your body to form carbon dioxide. Chemists describe these changes easily with *equations*.

$$\text{Magnesium} + \text{oxygen} = \text{magnesium oxide}$$
$$\text{Iron} + \text{oxygen} = \text{iron oxide}$$
$$\text{Carbon} + \text{oxygen} = \text{carbon dioxide}$$

The oxidation of magnesium

Place about 15 cm of magnesium ribbon inside a crucible. Partly cover the crucible with its lid. Weigh the crucible. Heat the crucible until the magnesium begins to burn. Reweigh the crucible when the magnesium has finished burning. It has increased in weight. The magnesium oxide formed weighs more than the original magnesium. Examine the magnesium oxide inside the crucible. It is a white powder. Note: do *not* gaze at the burning magnesium. The bright flame of burning magnesium can be harmful to the eyes.

Air is necessary for rusting

Fill a small flask with water. Boil the water to expel all the air. Place a clean iron strip inside the flask. Seal the flask with a rubber stopper. Place an iron strip in another flask containing unboiled water. This flask is not stoppered. Allow the flasks to stand for a few days. The iron in the second flask rusts. The iron in the first flask does not rust; there is no air available.

Rust prevention

Stainless steel cutlery and tools contain the metal *chromium*, which does not rust. "Tins" are wrongly named, for they are not made of tin, but of tinplate, which is prepared by dipping sheets of iron in molten tin. The thin layers of tin do not allow air and water to reach the iron. The iron used for outdoor work is *galvanized* by dipping it in molten *zinc*. Zinc does not rust. Tools and machinery are often rubbed over with an oily cloth. The thin layer of oil helps to keep out air and moisture. Paint is also a protection against rust.

Making oxides

Your teacher should do the following experiment for you; it could be dangerous. Lower a deflagrating spoon containing heated *sodium* into a gas-jar of oxygen. The sodium burns rapidly. Remove the spoon, and then seal the jar with a grease-plate. Repeat with *potassium, calcium,* sulphur, phosphorus and charcoal.

Kinds of oxides

Oxides are *acidic, basic* or *neutral*. Usually the oxides of non-metals are acidic. When they are *dissolved* in water they turn blue *litmus* red. Some oxides of metals are *basic*. When they are dissolved in water they form *alkalis* and turn red litmus blue. Neutral oxides do not have these effects.

Litmus is a dye which changes colour when it is in contact with acids and alkalis.

Testing oxides

Put a little water in each of the gas-jars used in making oxides. Shake them vigorously so that the oxides are dissolved. Dip two pieces of litmus paper, one red and one blue, into the water in each jar. The blue litmus turns red in the jars containing the acidic oxides of sulphur and phosphorus. The red litmus turns blue in the jars containing the alkaline oxides of the metals.

Bleaching with sulphur dioxide

Hang a coloured flower inside a bell-jar. Hold the jar over a dish containing burning sulphur. *Sulphur dioxide* gas enters the jar and the flower becomes white in colour. Repeat with straw and pieces of cloth.

Sulphur dioxide as a fumigant

As well as being used as a *bleaching* agent, sulphur dioxide is used as a *fumigant*. It kills some of the germs which cause diseases.

OXIDATION

Material + oxygen = oxide

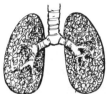

Carbon + oxygen
= carbon dioxide

BREATHING

Iron + oxygen
= iron oxide

RUSTING

Magnesium + oxygen
= magnesium oxide

BURNING

THE OXIDATION OF MAGNESIUM

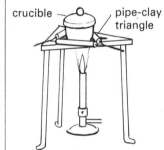

crucible — pipe-clay triangle

Oxidation — lid

Increased in weight — oxide

1 Weigh 2 Burn 3 Reweigh

Note the increase in weight

AIR AND RUSTING

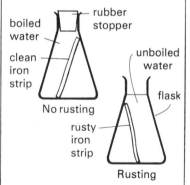

boiled water
clean iron strip
No rusting

rubber stopper
unboiled water
flask
rusty iron strip
Rusting

RUST PREVENTION

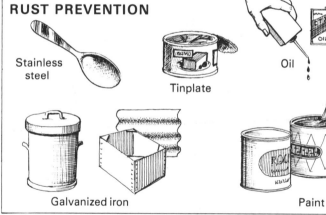

Stainless steel

Tinplate

Oil

Galvanized iron

Paint

MAKING OXIDES

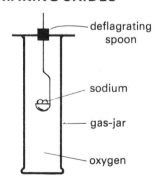

- deflagrating spoon
- sodium
- gas-jar
- oxygen

TESTING OXIDES

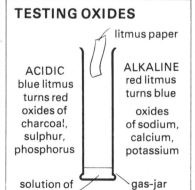

litmus paper

ACIDIC
blue litmus turns red
oxides of charcoal, sulphur, phosphorus

ALKALINE
red litmus turns blue
oxides of sodium, calcium, potassium

solution of oxide

gas-jar

BLEACHING

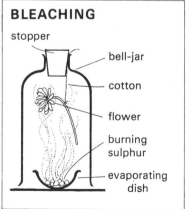

stopper

- bell-jar
- cotton
- flower
- burning sulphur
- evaporating dish

14   *Acids and Alkalis*

Acids

Vinegar and fruit juices are acids. They taste sour and turn blue litmus red. **Many acids are poisonous.** The *mineral acids, hydrochloric, sulphuric* and *nitric,* are poisonous, and they should be neither touched nor tasted. The fumes from these acids are also poisonous.

Alkalis

Alkalis turn red litmus blue and are often corrosive. *Caustic soda, caustic potash* and *ammonia* (smelling salts) are alkalis. **Do not touch alkalis.** Caustic soda and caustic potash dissolve animal and vegetable materials. You can imagine what effect these alkalis would have on your skin.

Testing with litmus

Place small quantities of the following materials, in turn, in a dish and test with litmus, washing the dish after each test. Use two pieces of litmus paper, one blue and one red, for each test. Notice which paper changes colour. Make two lists of acid and alkaline substances. Solids must be dissolved in water.

Vinegar; dilute nitric acid; dilute hydrochloric acid; sodium carbonate (washing-soda); sodium bicarbonate (baking-soda); gooseberry juice; sour milk; onion juice; juice of orange; sour beer; juice of lemon; weak ammonia solution; citric acid; sour-apple juice; blackcurrant juice; rhubarb juice; dilute caustic soda solution; dilute caustic potash solution; lime-water; soap solution; solutions of washing-powders.

Salts

When an acid and a *base* are mixed together, a *salt* is formed. Water is also formed. A base is an oxide of a metal. If just the right amounts of hydrochloric acid and caustic soda are mixed together in water, table salt, or sodium chloride, is formed. A chemist would say that an acid is *neutralized* by a base to make a salt and water.

Some common salts are *copper sulphate, sodium sulphate* (Glauber's salt) and *magnesium sulphate* (Epsom salt). They can be made by the action of certain acids on certain bases.

Acid + base = salt + water
Hydrochloric acid + caustic soda = sodium chloride + water
Sulphuric acid + *copper oxide* = copper sulphate + water

All acids contain hydrogen and all bases contain metals. All alkalis are bases. They are hydroxides of metals. Hydrochloric acid and sulphuric acid are sometimes written as hydrogen chloride and hydrogen sulphate. Sodium and copper are metals.

Treating stings

The poison in a bee's sting is acidic. Therefore, a bee sting should be treated with a mild alkali, such as ammonia or wet washing-soda. The poisonous acid is neutralized. The poison in a wasp's sting is alkaline. Therefore, a wasp sting should be treated with a mild acid, such as vinegar or the juice of a lemon.

Making table salt

Slowly pour dilute hydrochloric acid into an evaporating dish containing dilute caustic soda solution until a piece of blue litmus paper in the dish turns red. Stir the mixture. Add a few drops of caustic soda solution until the litmus paper takes on a purple colour—which is intermediate between red and blue. This means that the liquid in the dish is a neutral solution—the alkali has been neutralized by the acid. Heat the dish over a bunsen flame until all the water in it has been removed by evaporation. The dish now contains a white deposit of table salt. Taste this salt. A perfectly harmless salt has been made from two poisonous substances.

Note: not all salts are harmless. Copper sulphate is poisonous. Gardeners once used *fungicides* that contained copper sulphate and other poisonous salts to destroy moulds and bacteria.

ACIDS AND ALKALIS

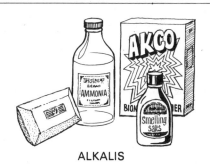

ACIDS

Vinegar
Dilute hydrochloric acid
Dilute nitric acid
Dilute sulphuric acid
Gooseberry juice
Sour milk
Onion juice

Sour beer
Orange juice
Lemon juice
Citric acid
Sour-apple juice
Blackcurrant juice
Rhubarb juice

turn blue litmus
red

ALKALIS

Washing-soda solution
Baking-soda solution
Dilute caustic soda solution
Dilute caustic potash solution

Ammonia solution
Lime-water
Soap solution
Solutions of
washing-powders

turn red litmus
blue

TESTING WITH LITMUS

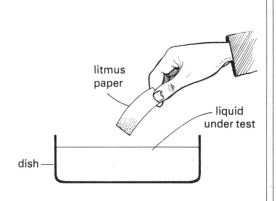

litmus paper

liquid under test

dish

TREATING STINGS

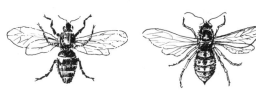

BEES

Treat
with
mild
alkalis

WASPS

Treat
with
mild
acids

MAKING TABLE SALT

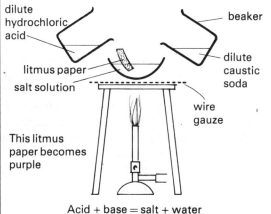

dilute hydrochloric acid

beaker

litmus paper

dilute caustic soda

salt solution

wire gauze

This litmus paper becomes purple

Acid + base = salt + water

DANGER

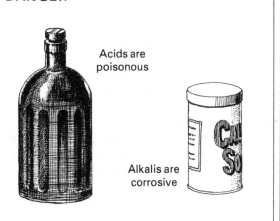

Acids are poisonous

Alkalis are corrosive

The uses of acids

Acids are used in car batteries, fire extinguishers and fruit salts, for cleaning and engraving metals, and in making hydrogen and other gases. They have many other uses.

Making fruit salts

Mix together by mass 7 parts of baking-soda (sodium bicarbonate), 6 parts of *tartaric acid*, 2 parts of pure *citric acid* and 3 parts of sugar. Place a tablespoonful of this mixture in a tumbler of water and stir it. The salts fizz in the water. Drink some of the liquid in the tumbler. The baking-soda, which is mildly alkaline, is neutralized by the tartaric and citric acids to give salts, water and carbon dioxide gas. The bubbling caused by the carbon dioxide gas is called *effervescence*. The citric acid and sugar give the drink its sweet, lemon taste.

A model fire extinguisher

Half fill a conical flask with a solution of baking-soda. Carefully suspend a small test-tube containing strong hydrochloric acid inside the flask. Insert a rubber stopper with a glass jet into the neck of the flask. Shake the flask. The acid is thrown out of the test-tube and effervescing liquid leaves the jet with some force. Fires will not burn in carbon dioxide gas.

> Hydrochloric acid + *sodium bicarbonate* (baking-soda) = sodium chloride + water + carbon dioxide

Hydrogen

Acids contain hydrogen. Hydrogen is the lightest gas known and it burns explosively in air. It is colourless and has no smell. It is used for filling balloons and hardening fats and oils. Water consists of hydrogen and oxygen.

Making hydrogen

Hydrogen can be prepared in the laboratory by the action of acids or water on certain metals. Here are three ways of making hydrogen:

Place some magnesium ribbon in a large beaker, and then partly fill the beaker with water and stand an inverted funnel in it so that the funnel covers the magnesium. Fill a test-tube with water and cover its mouth with your thumb. Invert the test-tube and, holding it under the water, remove your thumb. Bring the test-tube down over the end of the funnel tube. The magnesium is slowly oxidized by the oxygen in the water, and hydrogen gas collects in the test-tube.

> Water + magnesium = magnesium oxide + hydrogen

Place a few pieces of *granulated zinc* in a beaker of water. Place a test-tube, filled with water, over the zinc (see the previous experiment). Now add a little dilute hydrochloric acid to the water in the beaker. Hydrogen collects in the test-tube. Test for hydrogen with a lighted taper, holding the test-tube well away from your face. Bring the taper near to the open end of the test-tube. The hydrogen burns explosively with a *pop*.

> Zinc + hydrochloric acid = zinc chloride + hydrogen

Your teacher should do the following experiment for you because it could be dangerous. Drop a small piece of the metal sodium, about the size of a pea, into a dish containing water. Hydrogen gas is formed. The gas burns on the surface of the water. Take care with this experiment, for pieces of hot sodium may be thrown out of the dish.

> Sodium + water = sodium hydroxide (caustic soda) + hydrogen

Water from hydrogen and oxygen

Fill a large can with cold water. Suspend the can from a stand. Place a dish below it. Allow a small blue bunsen burner flame to play on the base of the can. The hydrogen in the gas combines with oxygen in the atmosphere and water vapour is formed. This condenses on the bottom of the can and drips off into the dish. Test for water with blue cobalt chloride paper (see page 18, Book One).

> Hydrogen + oxygen = water

MAKING FRUIT SALTS

Tartaric acid	—6 parts
Citric acid	—2 parts
Sugar	—3 parts
Baking-soda	—7 parts

SODA-ACID FIRE EXTINGUISHER (section)

- nozzle
- seal
- solution of baking-soda
- handle
- bottle of sulphuric acid
- small hole for escape of acid
- knob
- metal spike

MODEL FIRE EXTINGUISHER

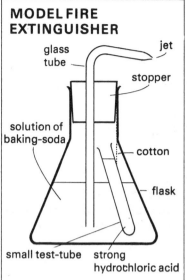

- glass tube
- jet
- stopper
- solution of baking-soda
- cotton
- flask
- small test-tube
- strong hydrochloric acid

MAKING HYDROGEN

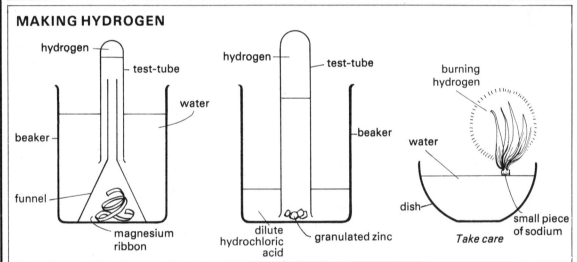

- hydrogen
- test-tube
- water
- beaker
- funnel
- magnesium ribbon
- hydrogen
- test-tube
- beaker
- dilute hydrochloric acid
- granulated zinc
- burning hydrogen
- water
- dish
- small piece of sodium
- *Take care*

HYDROGEN

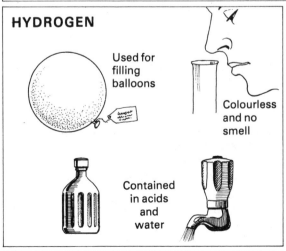

Used for filling balloons

Colourless and no smell

Contained in acids and water

WATER FROM HYDROGEN AND OXYGEN

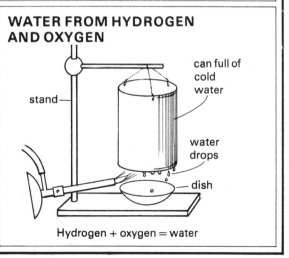

- can full of cold water
- stand
- water drops
- dish

Hydrogen + oxygen = water

Solutions

When salt is placed in water and stirred, it disappears. The salt *dissolves* in the water. It is still there even though it cannot be seen. Water with salt dissolved in it is called a *solution*. The liquid part of a solution is called the *solvent*. The solid matter which dissolves in it is called the *solute*.

Some materials do not dissolve in water. These are called *insoluble substances*. Wood, for example, is not soluble in water.

Some common solvents

Substances which do not dissolve in water may dissolve in other liquids. Oil does not dissolve in water, but it dissolves in *turpentine*. Turpentine is a solvent for oil. For that reason turpentine (or turpentine substitute) is used for thinning paint. Paint contains oil. *French polish* is a solution of *shellac* in methylated spirit. *Tincture of iodine* is a solution of *iodine* in alcohol. *Benzine,* which dissolves grease, is used for removing stains from clothing.

Soluble and insoluble substances

Place a quantity of fresh *lime* in pure water in a bottle and shake it. Allow it to settle. Pour off a small quantity of the clear liquid into an evaporating dish and boil it. A deposit of lime in the dish shows that lime is soluble in water. Repeat this experiment with each of the following substances. Make a list of those which are soluble in water.

Marble chips; table salt; sugar, alum; shellac; paraffin wax; clean sand; iron filings; borax; sodium carbonate; baking-soda; Epsom salt.

Salt in sea-water

Boil a quantity of sea-water in an evaporating dish until all the water has been lost by evaporation. A white deposit, which is a mixture of salts, remains in the dish. Place a small quantity of this white deposit on your tongue. It tastes salty.

Solutions in nature

Plants feed on the mineral salts dissolved in soil water. Mineral salts in the ground are dissolved by rain water. Rivers carry these solutions to the sea. This is one of the reasons why the sea is *saline*.

Saturated and unsaturated solutions

A solution in which no more of a solute will dissolve is said to be *saturated*. A solution which is capable of dissolving more of a solute is said to be *unsaturated*. Some substances are more soluble than others. Photographers' *hypo (sodium thiosulphate)* is very soluble.

Dissolving hypo

Add some crystals of hypo to a small quantity of water in a test-tube until no more will dissolve. What do you notice?

The effect of heat

Add *alum* to water in a beaker until no more will dissolve. Heat a quantity of this saturated solution in another beaker. Add a little more alum. It is dissolved. Its solubility has been increased by the rise in temperature.

The solubility of a substance in a liquid usually increases with a rise in temperature.

Separation by filtration

Some mixtures can be separated by using a solvent.

Stir a mixture of iron filings and common salt in water. Pour the mixture into a filter paper in a funnel. The diagram shows how the filter paper is fitted into the funnel. Catch the salt solution, which is called the *filtrate,* in a beaker. The iron filings, which do not dissolve in water, are left on the filter paper. This deposit of iron filings is called the *residue*. Filter paper is porous and liquids are able to pass through it. Pour a little pure water into the funnel to remove all traces of salt. Recover the salt by evaporating the water from the solution in a dish.

Repeat this experiment with a mixture of sugar and sand.

SOLUBLE AND INSOLUBLE SUBSTANCES

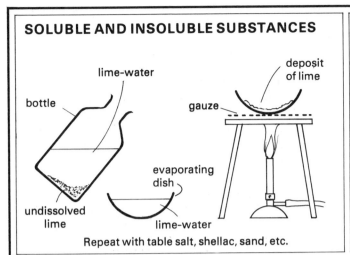

lime-water

deposit of lime

bottle

gauge

evaporating dish

lime-water

undissolved lime

Repeat with table salt, shellac, sand, etc.

SALT IN SEA-WATER

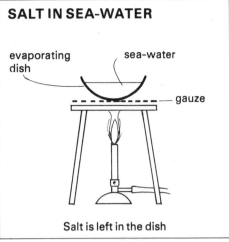

evaporating dish

sea-water

gauze

Salt is left in the dish

DISSOLVING HYPO

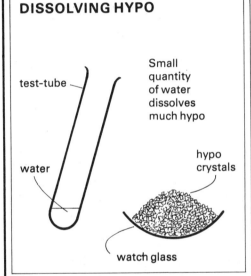

test-tube

Small quantity of water dissolves much hypo

water

hypo crystals

watch glass

THE EFFECT OF HEAT ON SOLUBILITY

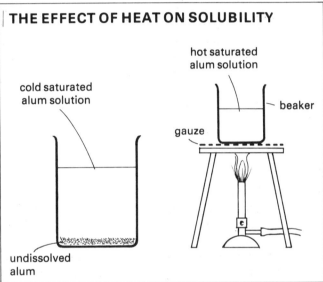

hot saturated alum solution

cold saturated alum solution

beaker

gauze

undissolved alum

SEPARATION OF A MIXTURE BY FILTRATION

FOLDING FILTER PAPER

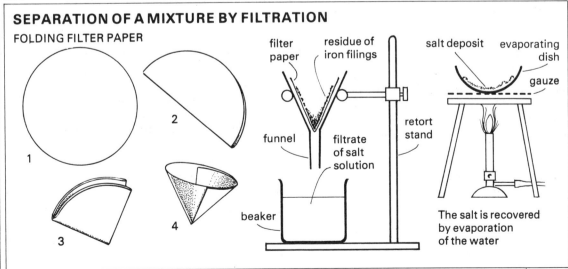

1

2

3

4

filter paper

residue of iron filings

funnel

filtrate of salt solution

beaker

retort stand

salt deposit

evaporating dish

gauze

The salt is recovered by evaporation of the water

17 | *Crystals*

Making copper sulphate crystals

Add copper sulphate to boiling water in a beaker. Continue to boil until no more of the salt dissolves. Discontinue boiling and allow the beaker to stand and cool. As the temperature falls, some of the solute cannot be held in solution, so crystals are formed. Dry the crystals with filter paper and examine them. Are they all the same shape?

Crystal shapes

Crystals of the same salt have the same shapes. Common salt crystals are rectangular blocks. Alum crystals are double pyramids. Soluble substances are usually crystalline. It is possible to identify substances by their crystals.

Substances are either crystalline or *amorphous*. Amorphous means "shapeless". Treacle and glass are amorphous substances. Do not make the mistake of thinking that glass is crystalline. Splinters of glass do not have exactly the same shape.

Water of crystallization

Some crystals contain water, which is known as *water of crystallization*. Salts from which the water has been removed are said to be *anhydrous*. Some salts absorb water vapour from the atmosphere, and they may even become liquid. These salts are said to be *deliquescent*. Salts which give off water to the atmosphere are said to be *efflorescent*. Table salt is deliquescent. Glauber's salt (sodium sulphate) is efflorescent.

Anhydrous copper sulphate

Strongly heat copper sulphate in a heat-resistant test-tube until a white powder of the anhydrous salt is formed. Allow it to cool. Add a little water to the salt. A blue copper sulphate solution is formed immediately. Anhydrous copper sulphate is used as a test for water.

Growing alum crystals

Suspend a small crystal of alum in a saturated solution of alum. Place a cover over it to keep out dust. Allow the solution to stand for several weeks. As evaporation occurs a large crystal of alum grows.

Making a crystal star

Suspend a shape, such as a star, made of cotton-covered copper wire, in a saturated solution of alum. After a few days, the wire will be covered with crystals. The crystals grow better on a rough surface.

Making a crystal "garden"

Place clean sand in the bottom of a large glass jar. Fill the jar with a *water-glass* solution, made up of one part of water-glass and three parts of warm water. Colour the solution pale green by adding crystals of *potassium dichromate*. Now add a few crystals of the following substances. In a few days these will have grown into a "garden".

Copper sulphate; hypo; iron sulphate; potash alum; chrome alum; magnesium sulphate; cobalt chloride.

Making bath salts

Put about 50 g of sodium carbonate into a large dish. Add 10 drops of a 1% dye solution. This solution is made by adding 1 ml of dye to 100 ml of water. Use a stick to mix the dye and the crystals thoroughly. Add a few drops of perfume and seal the crystals in stoppered bottles.

Stalactites and stalagmites

The formation of *stalactites* on the roofs and *stalagmites* on the floors of caves is due to deposits of *calcium carbonate* crystals being precipitated slowly from dripping water. Chalk and marble are forms of calcium carbonate.

Suspensions

Many, but not all, crystalline substances dissolve in water. Artificial solutions, called *suspensions*, can be made by shaking powders in water.

Making a suspension

Put a teaspoonful of chalk powder into a bottle half-filled with water. Shake the bottle vigorously. The water becomes milky in appearance. Allow the bottle to stand for several hours. The tiny particles of chalk slowly collect on the bottom of the bottle, and the water becomes clear.

CRYSTAL SHAPES

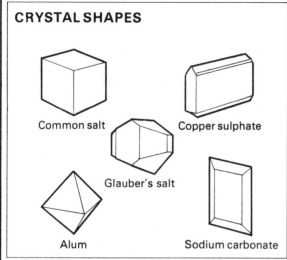

Common salt

Copper sulphate

Glauber's salt

Alum

Sodium carbonate

GROWING ALUM CRYSTALS

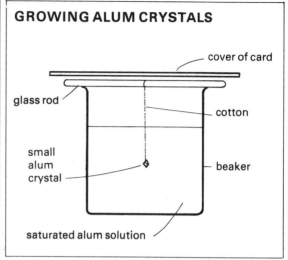

cover of card

glass rod

cotton

small alum crystal

beaker

saturated alum solution

CRYSTAL STAR

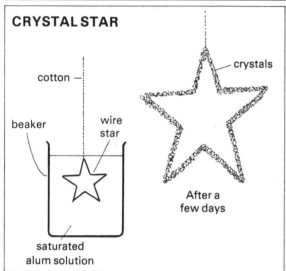

cotton

beaker

wire star

crystals

After a few days

saturated alum solution

CRYSTAL "GARDEN"

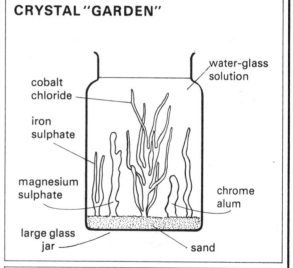

water-glass solution

cobalt chloride

iron sulphate

magnesium sulphate

chrome alum

large glass jar

sand

STALACTITES AND STALAGMITES

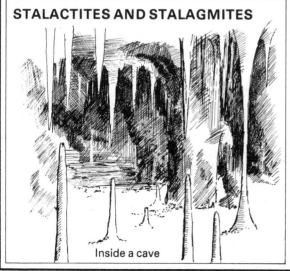

Inside a cave

MAKING A SUSPENSION

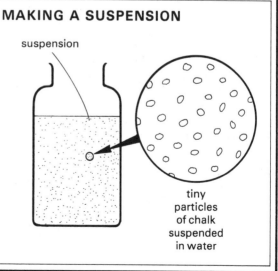

suspension

tiny particles of chalk suspended in water

General Science *Book Two*

The impurities in water

Fresh water from lakes and rivers is sometimes suitable for drinking when all the germs, dirt and rubbish have been removed. Sea-water is salty and is not used for drinking.

The mineral salts in water are sometimes beneficial. *Fluoride* salts help to prevent decay in teeth. Spa water, found in places like Harrogate and Droitwich, contains dissolved mineral salts that are said to be beneficial to people afflicted with rheumatic illnesses.

At a waterworks

Water is allowed to stand in reservoirs where rubbish can sink to the bottom. The oxygen in the atmosphere can reach the surface of the water and so kill many germs. Suspended particles of rubbish and some germs are removed by passing the water through *filter beds,* which are built up of layers of sand and gravel. Just sufficient quantities of the poisonous gas *chlorine* are added to the water to kill any remaining germs. This process is called *chlorination.*

Killing germs

The germs in water from springs and rivers can be killed by boiling it or adding small quantities of substances which give off oxygen, such as potassium permanganate and hydrogen peroxide. Campers sometimes do this.

A model water filter

Make a model water filter from two plant pots, clean sand, gravel, small stones and tiles. The diagram shows you how to do this. Pour some muddy water into the top of the filter. Clear water leaves the bottom. Do *not* drink the water for, though the mud has been removed, it still contains harmful germs.

Hard water

It is difficult to make a *lather* with water containing certain mineral salts. This water is called *hard water.* Water that contains small amounts of dissolved salts or no dissolved salts is called *soft water.* It is easy to make a good lather with rain-water. Most of the hardness in water can be removed by boiling it.

Kettle fur

The salts in hard water are deposited as a *fur* around the inside of a kettle. Less heat for boiling kettles is required if this fur, which is a poor heat conductor, is removed regularly.

Defurring a kettle

Half fill a large furred kettle with water to which vinegar has been added. Use 4 tablespoonfuls of vinegar to each litre of water. Allow the kettle to stand for four hours. Empty the kettle and tap out the fur. Fill the kettle with water, boil it, and then empty it. Rinse the kettle well.

Sometimes a marble or a small pebble is put in a kettle. During boiling, the marble moves about and helps to prevent fur from forming.

Testing for hardness

Add 15 g of soap shavings to 1 litre of *distilled* water and stir it. Fill a measuring jar with the soap solution. Put 100 ml of rain-water into a beaker. Pour a little of the soap solution into the rain-water and stir it vigorously. Continue to add soap solution until a lather is formed. Note the volume of soap solution that is required to make a lather that will last for a few minutes. Repeat this with river, sea, rain and distilled water. Which is the softest water. Which is the hardest? Why would it be difficult to wash in sea-water?

Making distilled water

Completely pure water is made by *distillation.* The apparatus shown can be used for this. The steam leaving the flask condenses on the cold walls of the test-tube. The test-tube is kept cool by a "jacket" of cold water. The impurities are left in the flask.

Make distilled water from a potassium permanganate solution. Taste the distilled water you have made. It tastes "flat".

THE IMPURITIES IN WATER

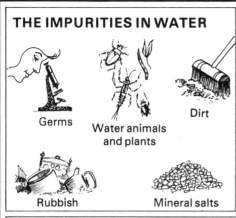

Germs

Water animals and plants

Dirt

Rubbish

Mineral salts

PURE WATER

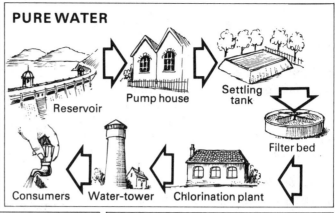

Reservoir

Pump house

Settling tank

Filter bed

Consumers

Water-tower

Chlorination plant

FILTER BED

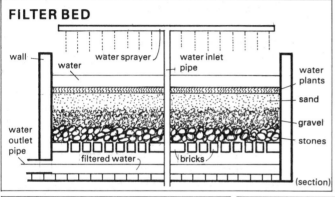

wall

water

water sprayer

water inlet pipe

water plants

sand

gravel

stones

water outlet pipe

filtered water

bricks

(section)

KILLING GERMS

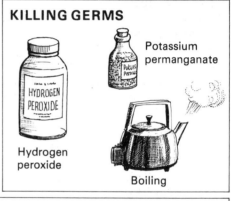

HYDROGEN PEROXIDE

Potassium permanganate

Potas Perma

Hydrogen peroxide

Boiling

MODEL WATER FILTER

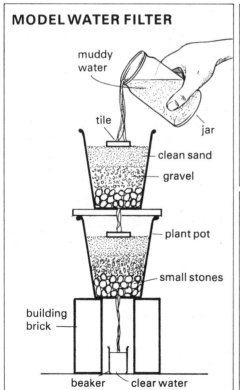

muddy water

tile

jar

clean sand

gravel

plant pot

small stones

building brick

beaker

clear water

TESTING FOR HARDNESS

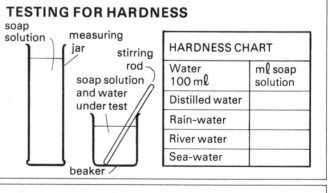

soap solution

measuring jar

stirring rod

soap solution and water under test

beaker

HARDNESS CHART	
Water 100 ml	ml soap solution
Distilled water	
Rain-water	
River water	
Sea-water	

MAKING DISTILLED WATER

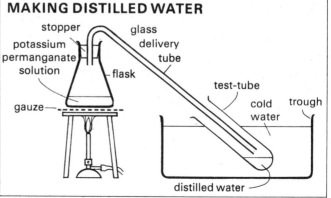

stopper

glass delivery tube

potassium permanganate solution

flask

gauze

test-tube

cold water

trough

distilled water

The "skin" on water

The surface of water acts like a thin skin which can support light objects that are denser than water. This "skin", which, of course, is not a real skin, is caused by forces in the water. It is called *surface tension*.

Surface tension

Sprinkle iron filings on the surface of some water in a dish. They float.

Then gently place a razor blade on the surface of the water. It floats.

Now place a small piece of filter paper on the surface of the water. Lay a needle on the paper. The paper becomes wet and then sinks. The needle remains on the surface, floating in a depression in the water.

More surface tension experiments

Float a needle on water. Then use a glass rod to put one drop of thin oil on the surface. The oil spreads and reaches the needle, which sinks. Do you know why?

Attach a small piece of *camphor* to the back of a model paper boat. Float the boat on water in a dish. What do you notice?

Bend about 30 cm of stiff wire to make a closed frame with a handle. Tie a loop of cotton across the frame. Dip it into a soap solution and lift out a soap film. Touch the film inside the loop with a hot needle. Surface tension causes the cotton to form a taut circle.

Pour water slowly and carefully into a jar until it rises just above the brim. Why does the water not spill? Why is its surface curved?

Liquids have surface skins

Soap bubbles and drops of water from a tap are round in shape. The particles in a drop of water are attracted to each other; those on the surface tend to be pulled inwards by those inside. The water drop becomes spherical and behaves as if it had an invisible skin. A large quantity of water has this "skin" too, but, because of its own weight, it is unable to become spherical. All liquids have surface tension. The surface tension of water is greater than that of oil or spirit.

Liquid drops are spherical

Allow water drops to fall into a tall glass jar which is full of linseed oil. The water drops are spherical because of surface tension.

Mix methylated spirit and water together in a tall glass jar. Suspend drops of oil in the mixture. The oil drops are spherical. Why?

Some insects use surface tension

Some insects, such as the whirligig beetle and the pond skater, walk and run on the surface of the water in a pond. The *larvae,* or grubs, of some insects, such as the gnat, hang downwards from the surface of the water. They are supported by surface tension.

When oil is sprayed on the water, the larvae are not supported because of reduced surface tension. Their breathing tubes are no longer exposed to the atmosphere and they die from lack of oxygen. In tropical countries, such as Panama, lakes and swamps are sprayed with oil in order to kill *mosquito* larvae. Mosquitoes carry the disease *malaria.*

Making an emulsion

Vigorously shake olive oil and water together in a bottle. Allow the bottle to stand. After a few hours, the mixture will have separated out into oil and water. Do this again, but, this time, add a small quantity of caustic soda. Examine the contents of the bottle after a few hours. The mixture has not separated. An *emulsion* has been formed. The soda, which is called an *emulsifying agent,* has broken down the oil into very tiny drops, which are suspended in the water.

Detergents

The grease on pots and pans is broken down by hot water and *detergents* to make an emulsion. It is difficult to wash greasy dishes in cold water without detergents.

SURFACE TENSION EXPERIMENTS

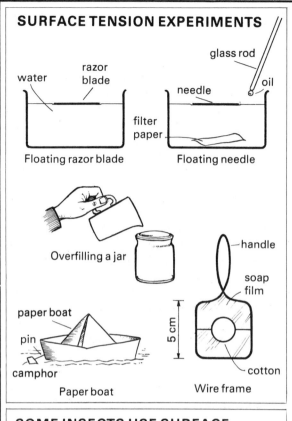

Floating razor blade

Floating needle

Overfilling a jar

Paper boat

Wire frame

SOME INSECTS USE SURFACE TENSION

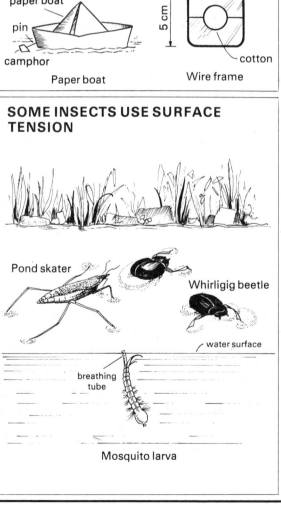

Pond skater

Whirligig beetle

water surface

breathing tube

Mosquito larva

THE "SKIN" ON WATER

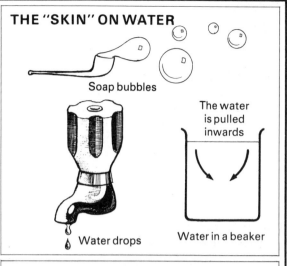

Soap bubbles

The water is pulled inwards

Water drops

Water in a beaker

LIQUID DROPS ARE SPHERICAL

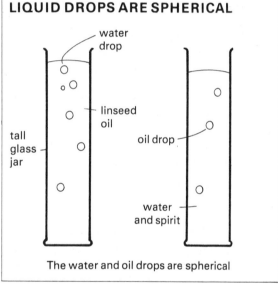

water drop

tall glass jar

linseed oil

oil drop

water and spirit

The water and oil drops are spherical

MAKING AN EMULSION

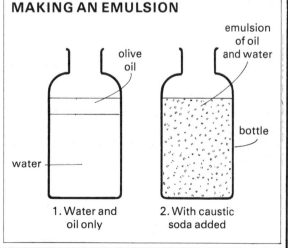

olive oil

emulsion of oil and water

bottle

water

1. Water and oil only

2. With caustic soda added

The meniscus on a liquid

Dip a glass rod into water. Remove the rod from the water. A thin film of water sticks to the rod.

Pour water into a narrow test-tube. Notice that the *meniscus* of the water is curved. Now pour mercury into another test-tube of the same width. What do you notice about the mercury meniscus?

Capillary attraction

A glass rod will attract and hold a thin film of water. A small drop of water hangs from a tap and does not fall. The water in a glass tube is attracted to its sides and a depression is formed. This force of attraction is called *capillary attraction*. This force overcomes surface tension. In the case of mercury, which is a very heavy liquid metal, capillary attraction does not overcome surface tension and the meniscus is curved upwards. Capillary attraction causes water to rise for some distance up thin glass tubes. Capillary attraction assists in the movement of the sap in plants.

Some capillarity experiments

Colour some water with red ink or dye in the proportions of 1 part ink and 4 parts water.

Stand various sizes of glass tubes with narrow bores in coloured water in a beaker. These narrow-bore tubes are called *capillary tubes*. Notice that the water rises most in the tube with the narrowest bore.

Place two glass plates together. Hold them in position with rubber bands. Insert a wooden splint between the plates at one side. The coloured water rises more on the side where the plates are closer together.

Place a small heap of sugar in a little coloured water in a dish. There is capillary attraction between the water and the sides of the sugar crystals. After a few seconds, the whole of the sugar is coloured.

Dip the end of a strip of filter paper into a dish containing coloured water. After a few seconds, the strip is completely coloured.

Plug one end of each of a few glass tubes, about 30 cm long, with cotton wool. Fill each tube with a sand of different coarseness. Stand the tubes in a dish containing coloured water that is a few centimetres in depth. The water rises most in the tube containing the finest sand.

Using capillarity

Capillarity is made use of in towels, pen-nibs, bath sponges, paintbrushes, blotting-paper and lamp and candle wicks. Towels are porous and soak up water. A towel made of a non-porous material, such as rubber, would not be very useful. Soil is porous. Water from low depths reaches the roots of plants by capillary attraction. The petrol in a lighter and the molten wax in a candle rise up the wicks by capillary attraction.

A model spirit-burner

Use a hammer and a nail to punch a hole, about 3 mm in diameter, in the centre of the screwed metal cap of a small glass jar. Use a thin nail to punch a small air-hole near the edge of the cap. Pour a quantity of methylated spirit into the jar. Replace the cap and insert a wick of cotton wool. Light the wick. Why is an air-hole necessary? Notice that, though the wick becomes slightly charred, it is the spirit which burns and not the wick.

Waterproofing

The natural oil on the feathers of water birds keeps out water and enables them to float. Builders provide houses with *damp courses* of lead, *bitumen,* plastic, slate or non-porous bricks so that water cannot rise up the porous walls. Polythene sheeting is widely used as a temporary cover during roof repairs. Mackintoshes, umbrellas and oilskins are made of non-porous materials. Outside woodwork is often *creosoted* so that water cannot enter the porous wood. Sloping roofs on buildings allow rain-water to run off easily.

THE MENISCUS ON A LIQUID

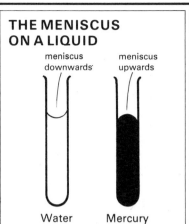

meniscus downwards

meniscus upwards

Water

Mercury

SOME CAPILLARITY EXPERIMENTS

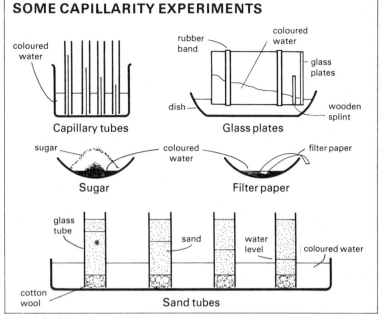

coloured water

Capillary tubes

rubber band

coloured water

glass plates

dish

wooden splint

Glass plates

sugar

coloured water

Sugar

filter paper

Filter paper

glass tube

sand

water level

coloured water

cotton wool

Sand tubes

USING CAPILLARITY

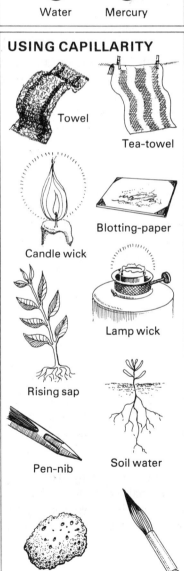

Towel

Tea-towel

Candle wick

Blotting-paper

Rising sap

Lamp wick

Pen-nib

Soil water

Bath sponge

Paintbrush

PETROL LIGHTER

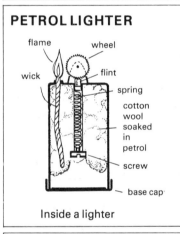

flame

wheel

wick

flint

spring

cotton wool soaked in petrol

screw

base cap

Inside a lighter

SPIRIT-BURNER

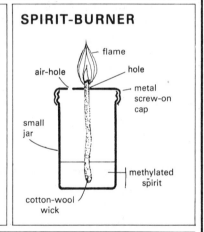

flame

air-hole

hole

metal screw-on cap

small jar

methylated spirit

cotton-wool wick

WATERPROOFING

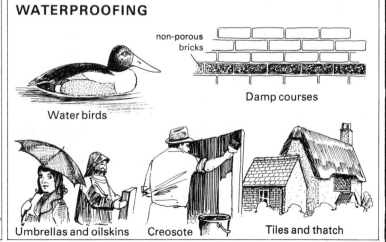

non-porous bricks

Damp courses

Water birds

Umbrellas and oilskins

Creosote

Tiles and thatch

Washing and Cleaning

Health and cleanliness

Oil and dirt carry germs. Some of these germs cause diseases. It is important, therefore, that you should protect your health by regularly removing dirt and oil from your body and your clothing.

Bath salts are made of sodium carbonate (washing-soda), dye and perfume. Dyes and perfumes give the salts an attractive appearance and a fragrant odour. Sodium carbonate is slightly alkaline. It is an emulsifying agent. It kills germs and emulsifies oil. It dissolves sweat, which sometimes smells unpleasant. It is fairly safe to use and does not injure the skin if it is not used excessively. It is a water softener.

Bath salts as a water softener

Add a small quantity of bath salts to some very hard water contained in a bottle. Shake the contents of the bottle vigorously. Allow the bottle to stand. A curd forms. This consists of the mineral salts which cause the water to be hard. Use soap to make lathers first with the hard water and then with the treated water. The treated water requires less soap to make a lather.

Soap

One of the first cleaning agents used by men was the wood ash taken from beneath cooking pots. This ash, which contains alkaline *potassium carbonate,* was called *potash.* Later on, *potassium hydroxide* was found to be more strongly alkaline, and, nowadays, it is often called *caustic potash.* By mixing caustic potash or caustic soda (*sodium hydroxide*) with animal fats, such as *tallow,* it was possible to make a soap that was not harmful to the skin. Tallow is animal fat.

Nowadays, soap is made from caustic soda and vegetable oils.

Potash is alkaline

Burn a few dry twigs in a metal dish. Shake the ash into a bottle containing a small quantity of water. Test the liquid with red litmus paper. The litmus paper turns blue showing that wood ashes contain an alkaline substance.

Making soap

Dissolve about 8 g of caustic soda in 100 cm^3 of distilled water contained in a large beaker. Add about 40 g of lard or a similar animal fat to the solution. Boil the mixture for a few hours and stir it frequently. Add water from time to time to replace the water lost by evaporation.

Prepare a saturated solution of common salt in 100 cm^3 of water. Pour the salt solution into the mixture and stir it well. Allow the resulting soap solution to stand and cool. The soap is "salted out" and rises to the surface.

Skim off the soap with a spoon. Put it into a small dish. The soap sets hard. Try it out by making a lather.

Soap and detergents

Toilet soap contains dye, perfume and good quality oil. Scrubbing soap is made from cheap fats. Carbolic soap contains antiseptic *carbolic acid.* Soap powders and detergents contain a water softener, like sodium carbonate or *borax.* Scouring soaps, used for removing thick grease and heavy dirt, contain an *abrasive,* like chalk or powdered *pumice.*

Making scouring soap

Mix together equal quantities of small soap shavings, powdered chalk and borax. Use enough of each substance to cover a 1p coin. Dab a wet cloth in the mixture and use it for cleaning taps, sinks, etc. in the laboratory. The chalk powder scours, or "scrapes off", grease and dirt.

Removing stains

Grease stains can be removed by using grease solvents, such as ammonia or methylated spirit.

Tar stains are removed in two stages. Soft fat is rubbed on the tar until it is all removed. Then the remains of the fat are removed with a grease solvent.

HEALTH AND CLEANLINESS

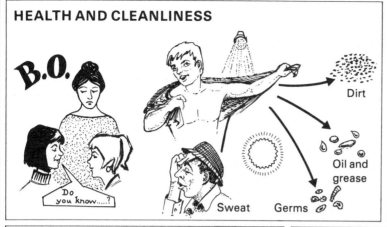

B.O.

Do you know.....?

Dirt

Oil and grease

Sweat   Germs

BATH SALTS AS A WATER SOFTENER

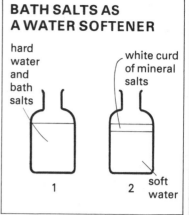

hard water and bath salts

white curd of mineral salts

1

2   soft water

SOAP

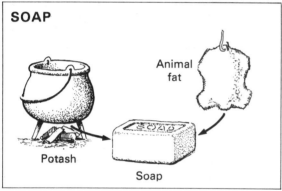

Animal fat

SOAP

Potash

Soap

POTASH IS ALKALINE

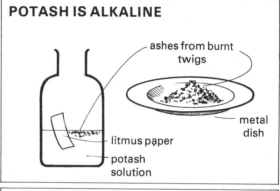

ashes from burnt twigs

metal dish

litmus paper

potash solution

MAKING SOAP

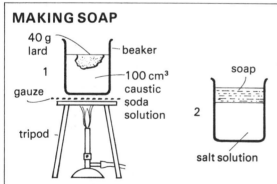

40 g lard

beaker

1

100 cm³ caustic soda solution

gauze

tripod

soap

2

salt solution

SCOURING SOAP

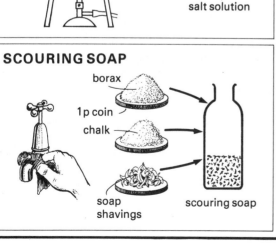

borax

1p coin

chalk

soap shavings

scouring soap

SOAP AND DETERGENTS

Toilet soap

Lavender

dye, good quality oils, perfume

Scrubbing soap

no perfume, no dye, cheap fat

Detergent

Carbolic soap

Carbolic

antiseptic

Soap powder

CAZ

water softener

liquid soap, water softener

Bath salts

washing soda, dye, perfume

Scouring soap

water softener, abrasive

Heat movement

Heat moves in solids and liquids by *conduction,* in liquids and gases by *convection,* and through space by *radiation.*

All matter is composed of tiny particles called *molecules.* In solid materials, heat moves from one particle to another in much the same way that a ball is passed from hand to hand along a line of rugby football players. This is conduction.

In liquids and gases, the warm particles rise. Warm water is not so heavy as cold water. This is convection. Convection can be likened to a player running with a ball.

Heat and light are able to move across space by radiation. The heat radiated by the Sun travels through 150 000 megametres of space before it reaches the Earth. Radiation can be likened to a player kicking or throwing a ball.

Using heat conductors

Generally, metals are good conductors of heat. Liquids are usually poor conductors. Mercury, which is a liquid metal, is an exception; it is a good heat conductor.

Metal objects feel colder than non-metal objects. Metals, being good conductors, carry warmth away from your fingers very quickly.

The bit of a soldering iron and the fire-tubes of a steam-engine are made of copper. Pans and kettles are often made of aluminium. Copper and aluminium are good conductors of heat. In an air-cooled engine, metal fins conduct the heat away.

Comparing heat conductors

Coat rods of different materials, such as wood, iron, copper, etc., 15 cm long and about 5 mm wide, with paraffin wax. Push the rods through corks. Insert the corks in holes in the lid of a flat metal can so that each rod extends 3 cm below the cork. Fill the can with hot water. Heat is conducted along the rods and the wax melts. The wax melts for the greatest distance along the best conductor.

Soak filter paper in a strong solution of cobalt chloride. Allow the paper to dry. Lay rods of different materials on pieces of the cobalt chloride paper. Heat the ends of the rods. When the rods become warm, the cobalt chloride papers change colour from pink to blue. The blueness extends for the greatest distance on the paper beneath the best conductor.

Water is a poor heat conductor

Strongly heat the top of a test-tube containing water. After a short time, carefully feel the bottom of the test-tube. The water at the bottom is quite cool, even though the water near the top is boiling.

The miner's safety lamp

Sir Humphrey Davy lessened the danger of gas explosions in coal-mines when he invented the miner's safety lamp. The flame of the lamp is covered by a wire gauze through which air reaches the flame. The heat from the flame is conducted away by the wires of the gauze, and the flammable gases outside the gauze are not heated by the flame and, therefore, they do not ignite.

Nowadays, mines are lit with electric lamps. However, the safety lamp is still used for the detection of poisonous gases. The flame changes colour in the presence of certain gases.

Examine a Davy lamp if you have an opportunity.

How the miner's safety lamp works

Make a coil by wrapping thick copper wire around a pencil. Bring the coil of copper wire down over a candle flame. The flame becomes smaller or is extinguished because its heat is rapidly conducted away by the copper wire.

Use tongs to hold a wire gauze over a bunsen burner flame. The flame does not pass through the gauze. Now turn on an unlighted burner under the wire gauze. Ignite the gas above the gauze. The gas below the gauze does not ignite. Why?

HEAT TRAVELS IN THREE WAYS

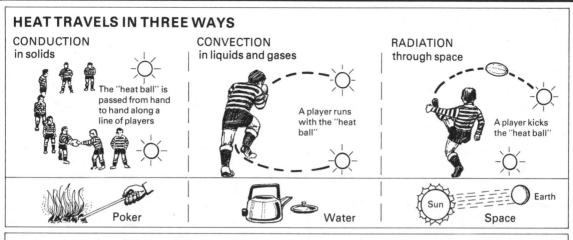

CONDUCTION
in solids

The "heat ball" is passed from hand to hand along a line of players

Poker

CONVECTION
in liquids and gases

A player runs with the "heat ball"

Water

RADIATION
through space

A player kicks the "heat ball"

Sun   Earth

Space

USING HEAT CONDUCTORS

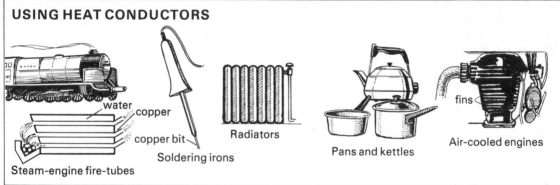

water
copper
copper bit

Steam-engine fire-tubes

Soldering irons

Radiators

Pans and kettles

fins

Air-cooled engines

COMPARING HEAT CONDUCTORS

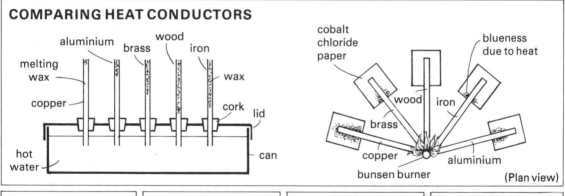

aluminium   brass   wood   iron

melting
wax

copper

wax

cork   lid

hot
water

can

cobalt
chloride
paper

blueness
due to heat

wood   iron

brass

copper   aluminium

bunsen burner

(Plan view)

WATER IS A POOR HEAT CONDUCTOR

boiling
water

test-
tube

cold
water

MINER'S SAFETY LAMP

handle

gauze

flame

container
for oil

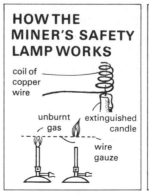

HOW THE MINER'S SAFETY LAMP WORKS

coil of
copper
wire

unburnt
gas

extinguished
candle

wire
gauze

SIR HUMPHREY DAVY

1778-1829

23 *Convection*

Convection

Water expands when it is heated. Therefore, a volume of warm water weighs less than an equal volume of cold water, and so warm water rises and cold water falls. These water movements are called *convection currents*. Air behaves in the same way when it is heated.

Hot water has less mass than cold water

Weigh a can full of cold water. Note its mass. Empty out the cold water and replace it with hot water. Re-weigh the can and its contents. The volume of hot water has less mass than the equal volume of cold water.

Some convection experiments

Put a few potassium permanganate crystals at the bottom of a large beaker of water. Heat the beaker over a small flame. The crystals dissolve slowly and colour the warm expanding water which is rising to the surface.

Drop some burning paper into a tall jar. The paper is extinguished because of lack of oxygen.

Again, drop burning paper into the jar, but, this time, slide a card down the middle of the jar. The paper now burns rapidly because of convection currents. Used air, which is warm, rises on one side of the card, and fresh air is drawn downwards on the other side of the card.

Heat the end of a poker until it is red-hot. Then, in a darkened room, use a torch to throw the shadow of the poker on to a screen. The shadows from the rising currents of warm air will be shown on the screen.

Make a paper saucepan (see the diagrams) and half fill it with water. Then heat the paper saucepan. The water boils but the paper does not burn. The heat which would burn the paper is carried away by convection currents in the water.

Two convection toys

Cut out a paper snake. Pivot it loosely on the pointed end of a bent wire. Hold the snake over a flame. It revolves. Why?

Cut out a paper fan, about 5 cm in diameter, with 8 vanes. Pivot it loosely on the end of a wire. Hold the fan over a flame. It revolves. Why?

Convection currents in nature

Perhaps you have noticed tiny specks of dust rising in a shaft of sunlight. They are carried upwards by currents of warm air.

Heat from the Sun helps to cause ocean currents. The water on the surface is heated and expands. It spreads outwards. Its place is taken by cold water from below the surface.

Winds are caused by convection currents. The air warmed by the Sun rises and cold air rushes in to take its place.

Land and sea breezes are also caused by convection currents. The pictures opposite show how breezes tend to blow from the sea during the day and from the land during the night.

Using convection

A chimney helps a fire to burn well. Hot air rises up the chimney and fresh air is drawn into the fire.

Convection currents help to ventilate rooms. Hot air rises up the chimneys and fresh air is drawn in through the doorways and windows.

Mines were once ventilated in a very dangerous way which often caused serious explosions. A fire was lit at the bottom of one shaft, which behaved as a chimney. Warm air rose up this shaft. Fresh air was drawn down another shaft.

Hot-water systems (of the kind described in Book One) and some central-heating systems depend for their operation on the convection currents in water.

The Romans had a form of central heating that was somewhat similar to the warm air under-floor heating systems in use today. The hot air from charcoal fires was drawn by convection currents through the foundations and chimneys of their buildings. In this way, the buildings were heated and smoke and fumes were kept out of the living quarters.

SOME CONVECTION EXPERIMENTS

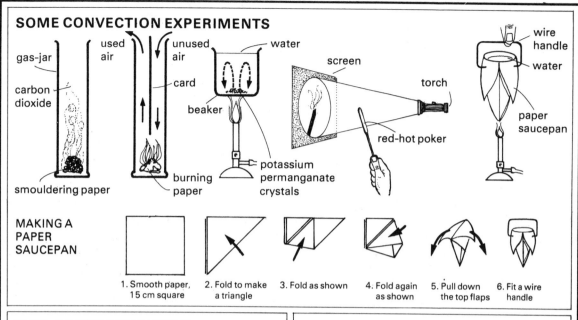

gas-jar
used air
unused air
water
screen
torch
wire handle
water
carbon dioxide
card
beaker
paper saucepan
red-hot poker
smouldering paper
burning paper
potassium permanganate crystals

MAKING A PAPER SAUCEPAN

1. Smooth paper, 15 cm square
2. Fold to make a triangle
3. Fold as shown
4. Fold again as shown
5. Pull down the top flaps
6. Fit a wire handle

TWO CONVECTION TOYS

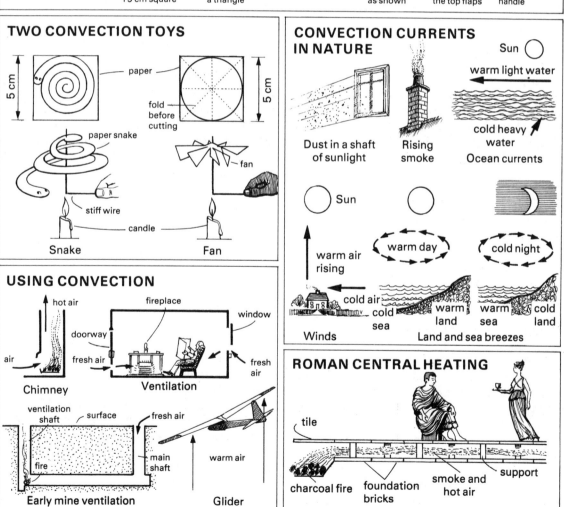

paper
5 cm
5 cm
fold before cutting
paper snake
fan
stiff wire
candle
Snake
Fan

USING CONVECTION

hot air
fireplace
window
air
doorway
fresh air
fresh air
Chimney
Ventilation

ventilation shaft
surface
fresh air
fire
main shaft
warm air
Early mine ventilation
Glider

CONVECTION CURRENTS IN NATURE

Sun
warm light water
cold heavy water
Dust in a shaft of sunlight
Rising smoke
Ocean currents

Sun
Sun
warm air rising
warm day
cold night
cold air
cold sea
warm land
warm sea
cold land
Winds
Land and sea breezes

ROMAN CENTRAL HEATING

tile
charcoal fire
foundation bricks
smoke and hot air
support

Radiation

When a pebble is dropped into a pond, ripples, or *waves,* travel outwards and across the surface of the water. In much the same way, heat and light waves are radiated through space from the Sun to the Earth. Light waves are radiated by a lamp. Radio waves are radiated by the mast of a radio transmitting station.

Heat radiation

Place your hand beneath an electric lamp which is on. Your hand feels warm. Convection currents of warm air rise upwards, and air is a poor conductor. Therefore, the heat must reach your hand by radiation. Switch off the lamp. Your hand now feels cool.

Heat absorption

Dark, unpolished surfaces absorb heat well. Bright, polished surfaces absorb heat badly. In tropical climates people sometimes wear white clothing because it absorbs little heat and so helps them to keep cool. Gardeners sometimes cover the soil with black polythene sheeting so that it will absorb heat. Plants grow better in warm soil than in cold soil.

Comparing surfaces as heat absorbers

Paint black one half of the inside of a large metal can. Leave the other half as it already is—bright and polished. Attach two coins with wax to opposite outsides of the can. Lower a white-hot or red-hot iron ball into the exact centre of the can. After a short time the wax melts and the coins fall. The coin on the black side is the first to fall because more heat is absorbed on that side than on the unpainted side.

Cover a thermometer bulb with black cloth. Cover another thermometer bulb with white cloth. Suspend a white-hot or red-hot iron ball midway between the two thermometer bulbs. The greater temperature rise is shown by the thermometer covered with black cloth. Why?

Sunshine and radiation

Heat from the Sun reaches the Earth by radiation. Some of this radiated heat is absorbed by the Earth's atmosphere. The atmosphere acts as a protective blanket. That is why, in bright sunlight, it is easy to become sunburnt on the top of a mountain; the atmosphere is thin at great heights.

Late frosts in April and May are caused when the heat absorbed by the earth during the day is radiated into space at night. Frosts are less likely to occur on a cloudy night, for clouds prevent the radiation of heat from the earth into space. Radiation causes the soil temperature to fall and dew is frozen into ice. Greenhouses and garden frames are covered with glass. Heat from the Sun passes through the glass by radiation and the air inside them becomes warm. This heated air cannot escape and so reaches a temperature which is much higher than that of the air outside.

How a garden frame works

Insert two thermometers, with their bulbs pointing upwards, into some garden soil on a sunny day. Cover one of the thermometers with a glass jar. Read the thermometers after a few hours. The thermometer covered by the jar has the higher temperature.

Heat reflection

Heat is reflected from a bright, polished surface in the same way that light is reflected by a mirror. Electric fires have polished reflectors, and radiators sometimes have silvered surfaces. Of course, the air around radiators is warmed mainly by convection currents.

A vacuum flask

One of the diagrams shows a *vacuum flask*. The silvered glass bottle inside the container has a double wall and is supported on pads of felt, cork or plastic. The air has been removed from the space between the walls of the bottle. Thus, heat cannot leave or enter the bottle by conduction or convection. The silvering prevents heat from leaving the bottle by radiation.

Dismantle an old flask. Examine its parts.

RADIATION

Waves   Light   Radio   Heat

HEAT RADIATION

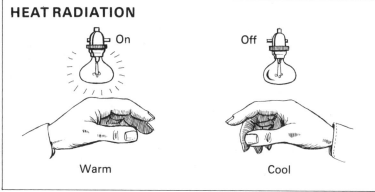

On   Off

Warm   Cool

COMPARING SURFACES

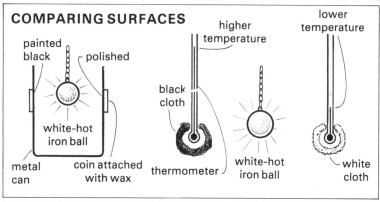

painted black   polished

higher temperature   lower temperature

black cloth

white-hot iron ball

metal can   coin attached with wax

thermometer   white-hot iron ball   white cloth

HOW A GARDEN FRAME WORKS

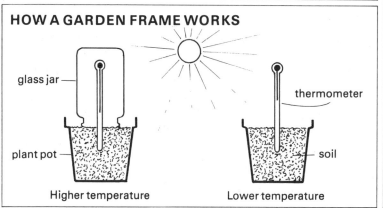

glass jar

thermometer

plant pot   soil

Higher temperature   Lower temperature

HEAT ABSORPTION

Bad absorbers   Good absorbers

Bright, polished surfaces   Dark, unpolished surfaces

Tropical clothing   Sheets of black polythene

SUNSHINE AND RADIATION

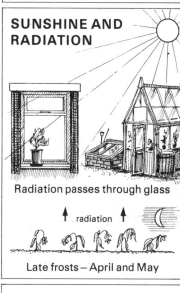

Radiation passes through glass

↑ radiation ↑

Late frosts – April and May

VACUUM FLASK

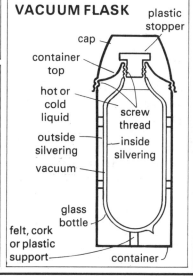

plastic stopper

cap

container top

hot or cold liquid

screw thread

outside silvering

inside silvering

vacuum

glass bottle

felt, cork or plastic support

container

The sky at night

When you look upwards on a clear night, you see many tiny stars, which seem to be scattered around the inside of a dark, bowl-shaped sky.

But the stars are not tiny. They are enormous in size and consist of gases at high temperatures. Our own Sun is small in comparison. They are separated from each other by immense distances. They seem to be small because they are so far away. The nearest star to the Earth, Proxima Centauri, is 40 000 *terametres* away. 40 terametres is 40 million million kilometres.

Movements of the stars are orderly. Hence, they have been used for checking dates in calendars. Ancient peoples used them for making prophecies. It was once believed that the stars controlled the destinies of men. Sailors and airmen can use them for checking their navigation—and desert peoples do so, too.

Astronomers are able, by means of large telescopes, to see some of the very distant stars which are not visible to the naked eye. But some stars are so distant that they cannot be investigated even with the most powerful radio telescopes.

There is an illustration opposite that shows a view of the northern sky (as seen by an observer in the Northern Hemisphere). You should notice the directions of east and west. If you hold the illustration above your head—where the real northern sky is—you will see that they are correct.

The constellations

Groups of stars are called *constellations*. "Stellar" means "star". By imagining lines joining stars, ancient peoples likened the constellations to objects, animals and people. The crossed lines of the constellation Cygnus have the appearance of a swan. *Cygnus* means "the Swan". Some constellations have the names of Greek and Roman gods and heroes. There is the constellation Hercules, for example.

Single stars have names, too. Sirius, the Dog Star, is the brightest star in the sky.

A constellation viewer

Make a card tube about 20 cm long and 4 cm in diameter. Use a needle to prick holes to represent a constellation on an 8 cm square of card. Make a card for each constellation. Hold the central part of a card against one end of the tube. Hold the other end against your eye and point the tube towards a bright light or a window. The constellation appears to you as it would do at night. You can use this viewer during the day, when the night sky is not available. You should look at the sky at night, however. Make a list of the constellations which you can recognise.

Finding direction

In the Northern Hemisphere, sailors once used the Plough, or Great Bear, for finding direction. The two end stars of this constellation, which are known as the Pointers, point towards the Pole Star, which is almost exactly above the Earth's North Pole. The Pole Star cannot be seen in the Southern Hemisphere, and so, in the Southern Hemisphere, the Southern Cross was used for finding direction. An imaginary line joining Crux to Acrux points towards the South Pole. The science of navigation by the stars is called *astronavigation*.

Finding latitude

Use a drawing-pin to attach a plumb-line, made from a thin thread and a lump of plasticine, and a card protractor to a wooden stand in the way shown opposite. (Many card protractors can be made if a celluloid protractor is used as a template.)

Use this *clinometer* to take a sight on the Pole Star. Measure the angle between the plumb-line which points to the Earth's centre, and the centre line of the protractor. This angle is your *latitude*.

Meteors and comets

A *meteor* is a small piece of material that is moving through space at a very high speed. It becomes *incandescent* when it travels through the Earth's atmosphere. Meteors are sometimes called *shooting stars*. A meteor which does not burn itself out completely falls on to the Earth. It is then called a *meteorite*.

A *comet* is a larger body which appears at regular intervals and moves in an elliptical *orbit* about the Sun. The "tail" always points away from the Sun.

VIEW OF THE NORTHERN SKY

The Pole Star is marked with a cross; the arrow passes through the Pointers.

Overhead

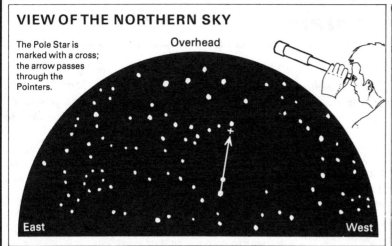

East

West

THE CONSTELLATION CYGNUS

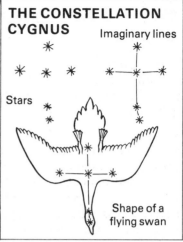

Imaginary lines

Stars

Shape of a flying swan

SOME CONSTELLATION SHAPES

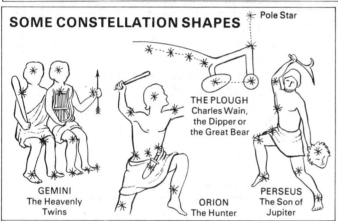

Pole Star

GEMINI
The Heavenly Twins

THE PLOUGH
Charles Wain, the Dipper or the Great Bear

ORION
The Hunter

PERSEUS
The Son of Jupiter

FINDING LATITUDE

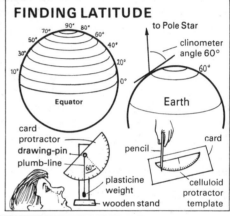

to Pole Star

clinometer angle 60°

Equator

Earth

card protractor
drawing-pin
plumb-line

pencil

card

plasticine weight

wooden stand

celluloid protractor template

SOME CONSTELLATIONS

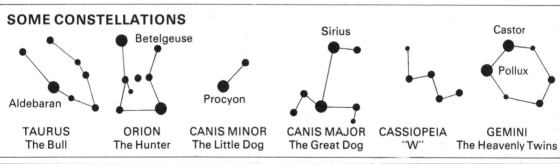

Betelgeuse

Sirius

Castor

Pollux

Aldebaran

Procyon

TAURUS
The Bull

ORION
The Hunter

CANIS MINOR
The Little Dog

CANIS MAJOR
The Great Dog

CASSIOPEIA
"W"

GEMINI
The Heavenly Twins

CONSTELLATION VIEWER

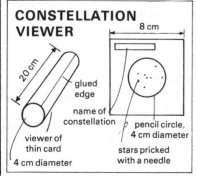

8 cm

20 cm

glued edge

name of constellation

pencil circle, 4 cm diameter

viewer of thin card
4 cm diameter

stars pricked with a needle

FINDING DIRECTION

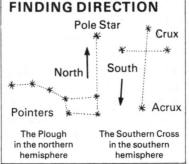

Pole Star

Crux

North

South

Pointers

Acrux

The Plough in the northern hemisphere

The Southern Cross in the southern hemisphere

A METEOR AND A COMET

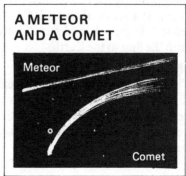

Meteor

Comet

The solar system

The Sun and its *planets* are called the *solar system*. The planets revolve around the Sun in regular *orbits*.

The planets

The planets of the solar system are Mercury, Venus, Earth, Mars, Jupiter, Saturn, Uranus, Neptune and Pluto. Between Mars and Jupiter, there is a ring of nearly 44 000 minor planets called *asteroids*. They may be parts from a large planet which broke up a long time ago. Jupiter and Saturn are both large planets. Jupiter has a ring of twelve moons and Saturn has nine. Two small planets are Mars and Mercury. Mercury is even smaller than either of two of Jupiter's moons. There is no life on the planets as we know it on the Earth. Mercury is too hot for life. Neptune and Pluto are too cold. Some of the planets have little or no atmosphere. Venus is seen in the east before sunrise. It is then called the Morning Star. When it is in the west, in the evening, it is called the Evening Star. The planets shine because they reflect light from the Sun.

A model solar system

Your classmates could represent the solar system by standing at the distances as given below. 1 cm represents 1 million kilometres.
From the Sun: Mercury 60 cm; Venus, 1 m; Earth, 1.5 m; Mars, 2.3 m; Jupiter, 7.7 m; Saturn, 14 m; Uranus, 28 m; Neptune, 45 m; Pluto, 60 m.

The Sun

The Sun, on which we depend for heat and light, is a small star. It is intensely hot. The metals in it exist as vapours. The distance between the Earth and the Sun is about 150 million kilometres. Light travels at a speed of about 300 000 kilometres per second. Light takes about 8.3 minutes to travel from the Sun to the Earth.

Night and day

The Sun appears to move from east to west. But, of course, it is the Earth which is moving from west to east. The side of the Earth which is facing the Sun is having day. The side which is in shadow is having night.

The Earth's rotation

Fill a large bowl with water. Sprinkle cork powder on the water. Make a line with powdered charcoal from the centre of the water surface to the side of the bowl. Make a mark on the bowl at the end of this line. Stand the bowl so that the line is in a north-south direction. After four hours you will notice that the line has apparently turned through an angle of about 60°. The bowl has moved with the Earth and the cork powder has remained still. In which direction does the line appear to move – east or west?

The Moon

The Moon is 400 000 kilometres away from the Earth. It is our nearest neighbour in space. Its surface is covered in volcanic craters. It is extremely hot during its day and extremely cold at night. It has no atmosphere, no water and no life. Astronauts who have visited the Moon have found it to be dry, dreary and altogether inhospitable.

Using a telescope

Look at the Moon through a telescope. The dark patches were once thought to be seas and oceans; they make the face of the "Man in the Moon".

The phases of the Moon

The Moon takes about 29 days, or a *lunar month,* to revolve once around the Earth. The *phases* of the Moon depend upon the amount of its illuminated surface which can be seen.

Eclipses

When the Moon is in the shadow of the Earth, it cannot be seen. This is an *eclipse* of the Moon. An eclipse or a partial eclipse of the Sun occurs when the Moon lies between the Earth and the Sun.

A Sun and Earth model

Use a globe, an electric lamp, a ball of plasticine and a knitting needle to make a model to show night and day, the Moon's phases and an eclipse of the Moon (see the diagrams).

THE SOLAR SYSTEM

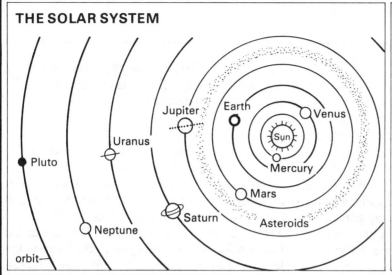

Jupiter
Earth
Venus
Sun
Mercury
Uranus
Pluto
Mars
Neptune
Saturn
Asteroids
orbit

THE PLANETS

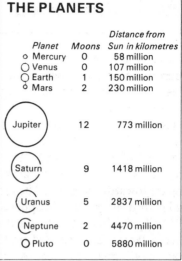

Planet	Moons	Distance from Sun in kilometres
Mercury	0	58 million
Venus	0	107 million
Earth	1	150 million
Mars	2	230 million
Jupiter	12	773 million
Saturn	9	1418 million
Uranus	5	2837 million
Neptune	2	4470 million
Pluto	0	5880 million

ECLIPSES

OF THE MOON   OF THE SUN

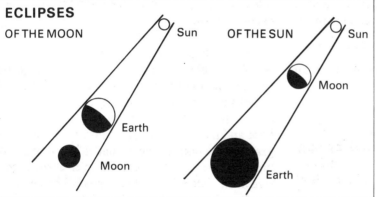

Sun
Earth
Moon

Sun
Moon
Earth

MODEL SOLAR SYSTEM

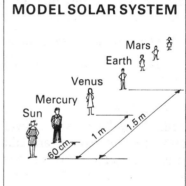

Mars
Earth
Venus
Mercury
Sun
60 cm
1 m
1.5 m

SUN AND EARTH MODEL

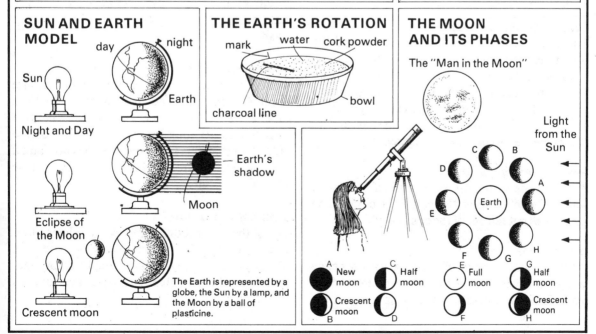

day   night

Sun
Earth

Night and Day

Earth's shadow

Moon

Eclipse of the Moon

Crescent moon

The Earth is represented by a globe, the Sun by a lamp, and the Moon by a ball of plasticine.

THE EARTH'S ROTATION

mark   water   cork powder

charcoal line   bowl

THE MOON AND ITS PHASES

The "Man in the Moon"

Light from the Sun

C   B
D   A
Earth
E
F   G   H

A New moon
C Half moon
E Full moon
G Half moon
B Crescent moon
D
F
H Crescent moon

27 Gravity

The force of gravity
Each object in the universe attracts every other object with a force that is known as the *force of gravity*. The pull of gravity depends upon the masses of the objects and the distances between them. The Sun's pull on the Earth is far greater than that of a distant star. Objects on the Earth do not fall off. The Earth's gravity tends to pull objects towards its centre. The force with which a ball is thrown upwards is opposed by gravity, so the ball falls back to the Earth. Large masses have a larger force exerted on them by gravity than have small masses. A volume of iron, therefore, weighs more than an equal volume of wood because the former has the greater mass.

Weight = mass × gravitational acceleration

Sir Isaac Newton studied gravity as the result of watching an apple fall to the ground.

Earth satellites
Artificial *satellites,* which revolve around the Earth like tiny moons, are launched by large rockets. They revolve in their orbits for some time until, on losing speed, they are pulled to the Earth by its force of gravity.

Centripetal and centrifugal forces
The Sun's gravitational force which keeps the planets in their orbits is called *centripetal force.* Planets are not drawn into the Sun by this force. This is because the planets are moving at high speeds. The reaction of the planets on the Sun is called *centrifugal force*, and is just enough to overcome the force of the Sun's gravity.

Showing centrifugal force
Attach string to the top of a can half filled with water. Rapidly swing the can in a circle. The water does not fall out of the can. Why?

A plumb-line
A plumb-line consists of a string with a heavy weight attached to it; the weight is often made of lead. The string points downwards towards the Earth's centre because of the force of gravity acting on the weight. Builders use plumb-lines for checking perpendiculars.

Tides
The Moon's gravitational pull on the Earth is the main cause of tides. The water on the Earth's surface is attracted by the Moon and builds up on opposite sides. At new moon, when the Sun and the Moon are exerting forces in the same direction, the tides are very high. They are called *spring tides*. At half moon, when the Sun and the Moon are exerting forces at right angles to each other, high tides are at their lowest. They are called *neap tides*.

The centre of gravity
An object behaves as if the Earth's force of gravity is acting upon one point within it. This point is called the object's *centre of gravity*. In a *stable* object, the centre of gravity is low down in its base and it does not fall over easily. The centre of gravity of a double-decker bus is low down. The bus is stable and does not fall over even when it sways on a steeply cambered road. The centre of gravity of an aeroplane is just above its wheels so that, on take-off, it does not overbalance and crash.

A balancing dancer
Make a skeleton of a dancer from a stiff thin wire about 60 cm long (see the diagram). Make a paper dress for the dancer. Then attach a lump of plasticine to her long leg. Now place her on the edge of a table. She rocks but she does not fall off the table. Her centre of gravity lies in or above the lump of plasticine.

Centre-of-gravity toys
One of the illustrations shows some other centre-of-gravity toys. Try making some of them. The Humpty Dumpty toy is made from an empty egg-shell, plasticine and glue. The toy sways but does not fall over.

THE FORCE OF GRAVITY

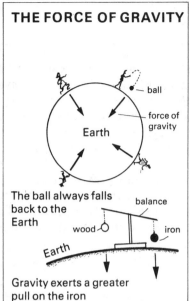

The ball always falls back to the Earth

Gravity exerts a greater pull on the iron

SIR ISAAC NEWTON
1642-1727

Mathematician and physicist. He saw an apple fall from a tree!

SHOWING CENTRIFUGAL FORCE

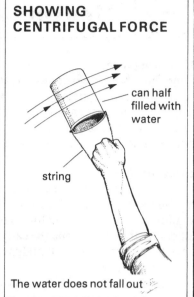

The water does not fall out

BUILDER'S PLUMB-LINE

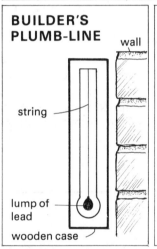

TIDES

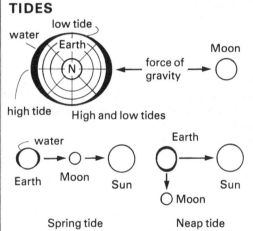

High and low tides

Spring tide   Neap tide

STABILITY

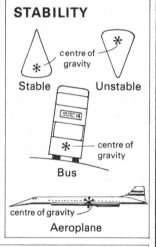

EARTH SATELLITES

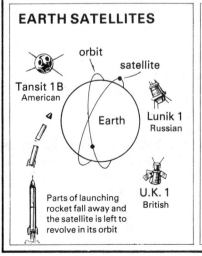

Parts of launching rocket fall away and the satellite is left to revolve in its orbit

BALANCING DANCER

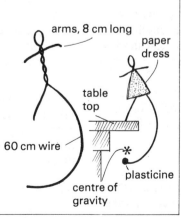

CENTRE-OF-GRAVITY TOYS

Measuring time

Time has been defined as "distance between events". Certainly, time can only be measured by regular events, such as night and day, the seasons, the phases of the Moon, the tides, the swings of pendulums in clocks, etc.

Using the Sun

Primitive people notice the seasonal changes and the Sun's movements. You could use your own shadow as a time-indicator. Your shadow is longest in the evening and the early morning. You have little or no shadow at midday, when the Sun is nearly overhead.

Cleopatra's Needle, a stone monument brought to London from Egypt and erected on the Thames Embankment, was used, about 1500 B.C., by the priests of Ancient Egypt as a kind of shadow clock.

The Earth's rotation

The Sun's apparent daily motion from east to west is due to the Earth's *rotation*. A *solar day* is the time between one noon and the next. A shadow cast by a stick is shortest at midday. In the Northern Hemisphere, the midday shadow points north. In the Southern Hemisphere, it points south.

The sundial is a type of shadow clock. The position of the shadow on the dial indicates the time. The sundial can be used only when the Sun is shining.

A shadow clock

Fix a broom handle upright in a playground on a sunny day. Use chalk to mark the length of its shadow. Do this every hour. The shortest shadow is cast at noon. Remember that, during winter in Britain, noon is at about 12 o'clock (G.M.T.—Greenwich Mean Time); during summer, noon is at about 1 o'clock (B.S.T.—British Summer Time). Clocks are put forward by one hour in spring and back by one hour in autumn.

The year and the seasons

A *solar year* is the time taken by the Earth to make one *revolution* around the Sun. It takes 365¼ days. For convenience, the calendar is arranged to contain 365 days. The quarter-days are added together to make one complete day, which is added on to February in a *leap-year*. A leap-year has 366 days. The Earth's axis is tilted. Therefore, days vary in their hours of daylight. In the Northern Hemisphere, the longest day is Midsummer Day, 21st June; the shortest day is 21st December. These days are known as the *solstices*. In the Southern Hemisphere, the longest day is 21st December and the shortest day is 21st June.

Time-measurers

The illustrations show some of the various devices which men have used for measuring time.

Making a candle clock

Use a ruler to measure the length of a candle. Burn the candle for 1 hour, then measure its new length. Subtraction gives the length of the candle which burns away in 1 hour. Now use a penknife to mark a candle with 1-hour, ½-hour and ¼-hour divisions.

Making a water clock

Make a water clock in the way shown opposite.

The pendulum

A *pendulum,* which is a string or a rod with a weight attached, is used for measuring small time intervals. A mechanical clock contains a pendulum. The *period,* or the time for the swing of a pendulum, depends upon its length. A pendulum always takes the same time to swing from one side to the other and back again.

Making a pendulum

Attach a plasticine bob to a 50 cm length of cotton. Suspend the pendulum. Tap it to set it in motion. Count the swings the pendulum makes in 30 seconds. Calculate its period.

$$\text{Period of pendulum} = \frac{30}{\text{no. of swings}} \text{ seconds}$$

USING THE SUN

The length of a shadow

Cleopatra's needle

SUNDIAL

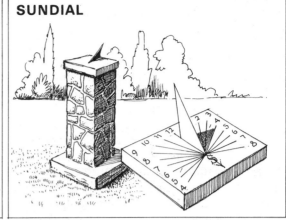

SHADOW CLOCK

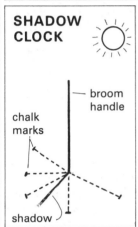

broom handle

chalk marks

shadow

THE SEASONS

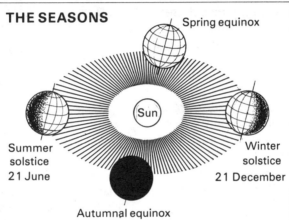

Spring equinox

Sun

Summer solstice 21 June

Winter solstice 21 December

Autumnal equinox

MAKING A PENDULUM

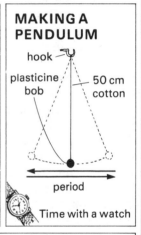

hook

plasticine bob

50 cm cotton

period

Time with a watch

MAKING A CANDLE CLOCK

Wax candle

Penknife

Ruler

MAKING A WATER CLOCK

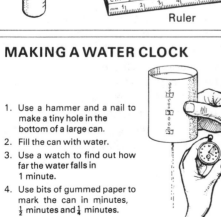

1. Use a hammer and a nail to make a tiny hole in the bottom of a large can.
2. Fill the can with water.
3. Use a watch to find out how far the water falls in 1 minute.
4. Use bits of gummed paper to mark the can in minutes, $\frac{1}{2}$ minutes and $\frac{1}{4}$ minutes.

TIME-MEASURERS

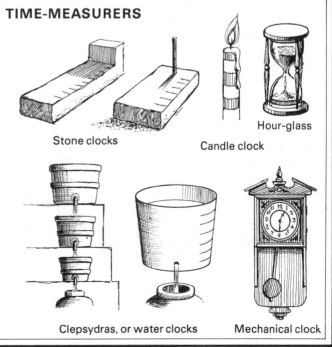

Stone clocks

Candle clock

Hour-glass

Clepsydras, or water clocks

Mechanical clock

The animal kingdom

All living things that are not plants are animals.

There are two main groups of animals—the *vertebrates* and the *invertebrates*.

Vertebrate animals have a *vertebral column,* or backbone. There are five kinds—*mammals, birds, reptiles, amphibians* and *fishes*. Dogs, rabbits, elephants, mice and humans are mammals. Snakes, tortoises and lizards are reptiles. Frogs and newts are amphibians. Crayfishes and lobsters, which do not possess backbones, are not true fishes even though they live in water. They are called *crustaceans*.

The invertebrates, which are sometimes called *non-vertebrates,* do not have backbones. Insects, spiders, molluscs and worms are invertebrates. Bees, wasps and houseflies are insects. Spiders, which are sometimes, quite wrongly, regarded as insects, have eight legs. Insects have only six legs. Molluscs are soft-bodied creatures. Most molluscs have shells. Garden snails, slugs, oysters, octopuses and mussels are molluscs. Worms are simple creatures with long bodies. The starfish and the jellyfish are very simple invertebrate animals. There are some tiny single-celled invertebrate animals, such as the *amoeba,* which cannot be seen with the naked eye.

A rabbit's skeleton

Examine a rabbit's skeleton. Notice that the vertebral column is made up of many small bones. These are called *vertebrae*. Draw a single vertebra.

The vertebrates

The table shows the main characteristics of the vertebrate animals.

Mammals have *mammary glands,* from which they supply their babies with milk. Humans, the most intelligent of the mammals, have established their supremacy over the other mammals and harnessed some of the forces of nature.

Mammals	Most have hair, fur or wool. Warm-blooded. Do not lay eggs. Supply their babies with milk. Well developed sense organs. Intelligence. Lungs.
Birds	Feathers. Warm-blooded. Most are capable of flight. Hard-shelled eggs. Well developed sense organs. Instinct. Lungs.
Reptiles	Mainly land animals. Cold-blooded. Some have round soft-shelled eggs. Some do not lay eggs. Lungs. Scales.
Amphibians	Cold-blooded. Many shell-less eggs. Intermediate stage between reptiles and fishes. Tadpoles live in water; adults can live on land. Gills and lungs.
Fishes	Cold-blooded. Many eggs. Live entirely in water. Fins. Gills. Scales.

Animal adaptation

Animals are wonderfully adapted to the conditions under which they live. Fishes, with their streamlined shapes, are ideally suited for life in water. Frogs have jointed legs and webbed feet, so they are as much at home in water as on land. Their long tongues enable them to capture the enormous quantities of flies that are an essential part of their diet. Giraffes have long necks, so they can eat the leaves on the higher branches of trees.

Outdoor observations

Here are some of the things you should look for when you are outdoors. The study of animals can become a most interesting hobby.

Activities: feeding, breathing, growth, movement, reproduction, care of the young, etc.

"Homes": burrows, nests, holes, shells, etc.

Migration: swallows, cuckoos, etc.

Hibernation: squirrels, bats, snails, etc.

Adaptation: fins, webbed feet, etc.

Defence: speed, size, strength, etc.

"Clothing": fur, hair, wool, feathers, scales, etc.

Variety: sizes, shapes, colours, etc.

THE ANIMAL KINGDOM

VERTEBRATES
Backbones

INVERTEBRATES
No backbones

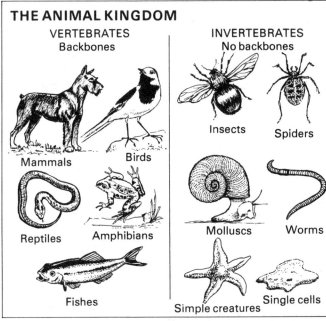

Mammals

Birds

Reptiles

Amphibians

Fishes

Insects

Spiders

Molluscs

Worms

Simple creatures

Single cells

RABBIT'S SKELETON

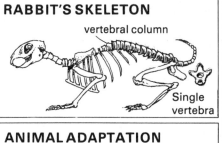

vertebral column

Single vertebra

ANIMAL ADAPTATION

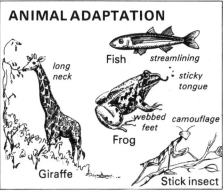

Fish

streamlining

long neck

sticky tongue

webbed feet

camouflage

Frog

Giraffe

Stick insect

ANIMAL ACTIVITIES

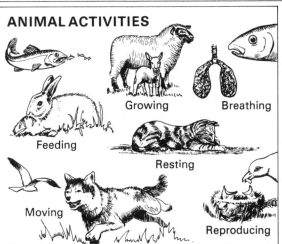

Growing

Breathing

Feeding

Resting

Moving

Reproducing

THE HOMES OF ANIMALS

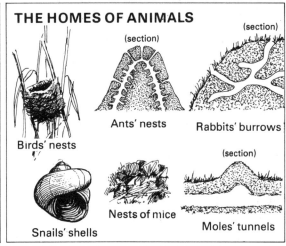

(section)

(section)

Ants' nests

Rabbits' burrows

Birds' nests

(section)

Snails' shells

Nests of mice

Moles' tunnels

HIBERNATION

Winter sleep of animals

ANIMAL DEFENCE

Speed

Size

Claws

Teeth

Hoofs

Sight

Horns

Smell

Poison

Beaks

Camouflage

Hearing

Mammal families

This table shows the main families of mammals with some typical examples.

Pouched	Kangaroo, koala bear
Insect-eating	Mole, hedgehog, shrew
Gnawing	Hare, rabbit, rat, mouse, squirrel
Elephants	Asiatic, African
Sea mammals	Whale, dolphin, porpoise, seal
Hoofed	Horse, rhinoceros, zebra, goat
Flesh-eating	Cat, lion, tiger, dog, fox, wolf
Manlike	Chimpanzee, gorilla, man

Mammals which feed on flesh are called *carnivores*. Those which feed on plants are called *herbivores*.

Some mammals, such as dogs, camels and horses, have been tamed and trained by men to do useful work. These are called *domesticated animals*.

Bats are the only mammals which can fly. Their wings are large membranes of skin stretched between their long fingers.

British mammals

Here are most of the wild mammals to be found in Britain:
Rabbit; hare; stoat; weasel; fox; rat; otter; seal; deer; hedgehog; mole; shrew; bat; mouse; squirrel; badger; vole.

Keeping mammals

White mice, rats, guinea-pigs, rabbits and hamsters can be kept in cages. Golden hamsters have no objectionable smell and are attractive.

One of the diagrams opposite shows how a wooden cage for hamsters can be constructed.

Feed the hamsters on table scraps, nuts, root vegetables, cabbage, seeds, grain, dry bread, etc. 30 g of food per day for each hamster is enough. Provide water and a "toilet" box. Sprinkle sawdust on the floor of the cage. This will absorb moisture. Keep the cage in a warm, shaded, draught-free position. Clean it out regularly.

Keep records of sizes, weights, colours, times of mating, times of births, numbers of young hamsters in litters, food, etc.

Beaks and feet

The beaks and feet of birds indicate their habits in feeding and movement. Eagles have hooked beaks and sharp claws which are useful for grasping and tearing flesh. Parrots' beaks are used for cracking nuts. The webbed feet of ducks are used in swimming. Herons have long pointed beaks for catching fish.

Observing birds

Observe birds and their activities when you are outdoors. Here are some of the things to look for: Shapes, sizes and materials of nests; colours and sizes of eggs; feeding; mating; differences between cock and hen birds; sizes, shapes and colours of birds; nestlings; fledglings; flight; beaks and feet; colonies; migration; song.

Do *not* interfere with nests, eggs or birds. Some birds are rare, and the wanton destruction of them is to be deplored. Acts of Parliament forbid the destruction of many species.

Making a bird-table

Make a bird-table out of a wooden board about 30 cm square. Nail laths on top of the board. Birds will perch on these. Place cold boiled potato, bits of fat, seeds, strings of peanuts, breadcrumbs and a dish of water on the table. Attach strings to its corners. Hang it up well out of the reach of cats.

Try to recognise the birds which visit the table. Write down their names. This is something which you can do at home.

Bird footprints

Put a smooth layer of moist soil on the top of a bird-table without laths. Notice the birds which come to the table. Pour plaster of Paris of the consistency of thick cream on their footprints. Remove the casts when they have set. Paint the impressions black. Label the casts with the names of the birds.

MAMMAL FAMILIES

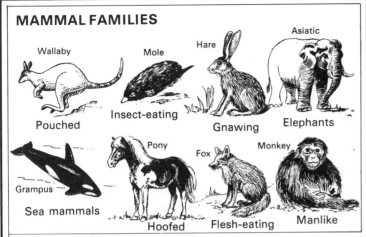

Wallaby — Pouched

Mole — Insect-eating

Hare — Gnawing

Asiatic — Elephants

Grampus — Sea mammals

Pony — Hoofed

Fox — Flesh-eating

Monkey — Manlike

SOME DOMESTICATED ANIMALS

Cat

Dog

Camel

Horse

Cow

KEEPING MAMMALS

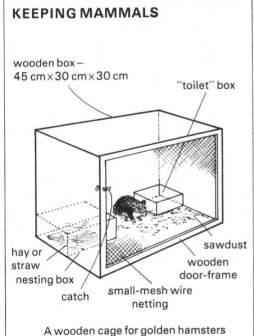

wooden box —
45 cm × 30 cm × 30 cm

"toilet" box

sawdust

wooden
door-frame

hay or
straw
nesting box

catch

small-mesh wire
netting

A wooden cage for golden hamsters

BAT

Die Fledermaus means
"the flying-mouse"

Not a bird but a mammal

SONG THRUSH

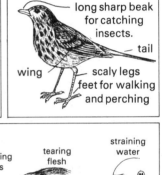

long sharp beak
for catching
insects.

tail

wing

scaly legs
feet for walking
and perching

BEAKS AND FEET

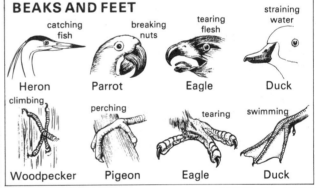

catching
fish — Heron

breaking
nuts — Parrot

tearing
flesh — Eagle

straining
water — Duck

climbing — Woodpecker

perching — Pigeon

tearing — Eagle

swimming — Duck

OBSERVING BIRDS

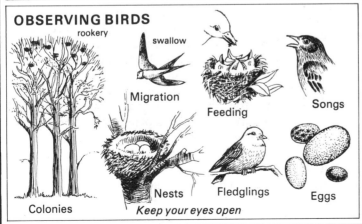

rookery

swallow

Migration

Feeding

Songs

Nests

Fledglings

Eggs

Colonies

Keep your eyes open

MAKING A BIRD-TABLE

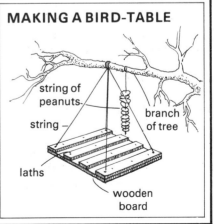

string of
peanuts

string

branch
of tree

laths

wooden
board

Cold-blooded Animals

Reptiles

Reptiles are cold-blooded animals. Strictly speaking, they should not be called cold-blooded animals, for their temperatures vary with the temperature of their surroundings. Mammals and birds are able to maintain a constant and high body temperature whether their surroundings are hot or cold. Snakes, crocodiles, alligators, lizards, tortoises and turtles are reptiles. Most reptiles are land animals, but some, such as the turtles and crocodiles, spend most of their lives in water.

British reptiles

There are only three kinds of snakes in Britain—the *grass-snake, adder* and *smooth snake*. The commonest is the grass-snake. It feeds on frogs and fishes and is quite harmless to humans. It is grey-green in colour. Adders, or *vipers,* are poisonous. An adder can be recognised by the black zigzag pattern along its brown back. Smooth snakes are rare.

There are only three kinds of lizards in Britain—the *common lizard, sand lizard* and *slow-worm*. The slow-worm is about 30 cm long. It has no limbs and looks like a very large earthworm.

Amphibians

The amphibians are cold-blooded animals too. They begin their lives in water and then change into land animals. Their skins are bare and not scaly. Frogs, toads, newts and salamanders are amphibians. Newts are similar in appearance to lizards but they are not reptiles. Though they have legs, they live in water. Frogs and toads are able to move about on land as well as in water. Toads have wrinkled and warty skins.

British amphibians

Newts, frogs and toads are the three kinds of amphibians found in Britain.

A frog

Examine a living or a preserved specimen of a frog. Notice its wide mouth, protruding eyes, exposed ear-drums, webbed feet with five toes on the hind legs, and webless feet with four toes on the front legs. Make a labelled drawing.

Metamorphosis

Newts, frogs and toads undergo a *metamorphosis,* or a change of body, during their lives. "Meta" means "change" and "morph" means "body". They begin life as eggs, which hatch to become tadpoles. These tadpoles have gills and live in water in the same way as fish. They grow legs and develop lungs. Then they are able to live on land in the same way as reptiles. It is interesting to note that the toad, which is a land animal, must begin its life in water as a tadpole.

Frogs lay masses of eggs which are surrounded by a jelly-like material. Toads lay their eggs in strings of jelly. Newts lay their eggs singly on the leaves of water plants.

The life history of a frog

One of the illustrations opposite shows the main stages in the life of a frog.

Keeping tadpoles

Put a quantity of frog *spawn* in a large glass tank containing water and water weeds. Frog spawn is found in March. Watch the development of the eggs. Put large stones in the trough so that the young frogs can leave the water and breathe with their lungs. Occasionally, the tadpoles can be fed on live water fleas and chopped meat, earthworms and boiled egg. But do *not* leave uneaten meat in the tank. It decays and fouls the water.

Making a vivarium

A *vivarium,* or a home for reptiles and amphibians, can be made from a large plastic bowl and a sheet of glass or transparent plastic. Fill the bottom of the bowl with living turf and a few stones. Provide dishes of water, sunk level with the turf, as accommodation for the newts and frogs. Snakes, lizards, newts, toads, frogs and tortoises can be kept in vivaria. Some of these animals must not be kept in the same vivarium. Snakes feed on frogs!

Feed the animals on insects, worms, chopped meat and boiled egg, etc.

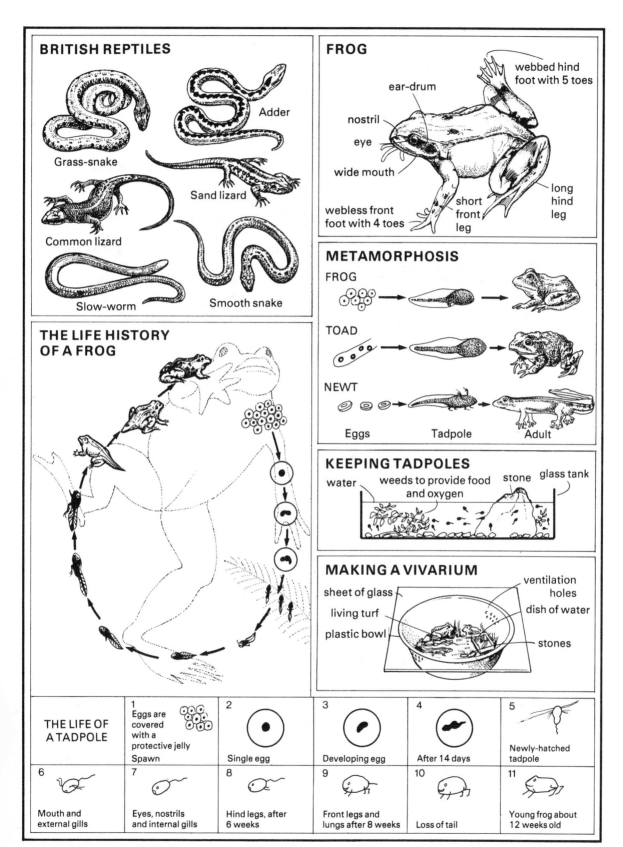

BRITISH REPTILES

Grass-snake

Adder

Sand lizard

Common lizard

Slow-worm

Smooth snake

FROG

webbed hind foot with 5 toes

ear-drum

nostril

eye

wide mouth

webless front foot with 4 toes

short front leg

long hind leg

THE LIFE HISTORY OF A FROG

METAMORPHOSIS

FROG

TOAD

NEWT

Eggs   Tadpole   Adult

KEEPING TADPOLES

water   weeds to provide food and oxygen   stone   glass tank

MAKING A VIVARIUM

sheet of glass

living turf

plastic bowl

ventilation holes

dish of water

stones

THE LIFE OF A TADPOLE	1 Eggs are covered with a protective jelly Spawn	2 Single egg	3 Developing egg	4 After 14 days	5 Newly-hatched tadpole
6 Mouth and external gills	7 Eyes, nostrils and internal gills	8 Hind legs, after 6 weeks	9 Front legs and lungs after 8 weeks	10 Loss of tail	11 Young frog about 12 weeks old

Insects

Insects make up the largest group of animals. Their bodies are divided into three parts—*head, thorax* and *abdomen*. The head has two jointed feelers and two large *compound* eyes. Three pairs of legs and, sometimes, two pairs of wings are attached to the thorax. One of the diagrams shows the general structure of an insect.

Insect families

Some common insect families are the beetles, moths, butterflies, ants, bees, wasps, *bugs,* flies and grasshoppers. In beetles, the first pair of wings is hard and acts as a protective cover. Grasshoppers have large back legs. They chirp by rubbing their hind legs against their wings. Flies have no second pair of wings. Houseflies and gnats are flies. Bugs are simple insects with their mouth-parts adapted for sucking juices. Lice and greenflies are bugs.

A housefly

Examine a housefly under a magnifying glass. Notice the three main parts of its body, its compound eyes, its six legs and its single pair of wings.

Metamorphosis

Many insects undergo a metamorphosis which is stranger even than that of amphibians. There are four stages in the life histories of these insects—*egg, larva, pupa* and *imago.* The larvae of butterflies and moths are called *caterpillars.* The larvae of flies are called *maggots.* The illustrations show the stages in the life history of a gnat, a housefly and a butterfly. The larvae lose their skins by *moulting* as they grow bigger. Another illustration shows a dragonfly with its *nymph.*

Keeping flies

Trap some blowflies in a bottle containing a little raw meat. Plug the neck of the bottle with cotton wool. Release the flies when their eggs have been laid. Watch the hatching and the growth of the maggots. Feed the adult flies on moist biscuit and sugar and water.

Spiders

Spiders are not insects. They belong to the *arachnid* family. A house-spider has eight legs, eight single eyes, and six *spinnerets,* or spinning tubes, under its abdomen. There are two parts to its body—abdomen and fused head and thorax. One of the diagrams shows the general structure of a spider.

A spider spins a web to trap insects. Flies are caught on the sticky strands of the web. The spider binds a fly with web, poisons it with its *fangs,* and then sucks out the soft parts. Spiders do not undergo a metamorphosis; baby spiders have the same structure as the adults but are smaller. Spiders wrap their eggs in protective *cocoons* of silk.

Scorpions and king-crabs belong to the arachnid family.

Keeping spiders

Put a few spiders in a large jar or insect cage. Place a few twigs inside the jar. The spiders will attach their webs to these. Feed the spiders on houseflies. Examine the webs which the spiders make.

Collecting insects

Collect insects when you are outdoors. The only equipment you need for this is a net and a metal can, the lid of which is punched with small air-holes. The net is used for catching flying insects. Your teacher and your classmates will help you to find out their names.

Keep the insects for a few days in large glass jars with lids, and then release them. Punch small air-holes in the lids. Insect larvae should be fed on the leaves of the plants on which they are found.

Preserve dead specimens by dipping them in molten paraffin wax. Attach them to cards with small pins or strong glue. Label the cards and cover them with transparent paper.

THE GENERAL STRUCTURE OF AN INSECT

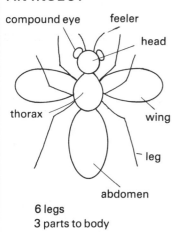

compound eye · feeler · head · thorax · wing · leg · abdomen

6 legs
3 parts to body

SOME INSECT FAMILIES

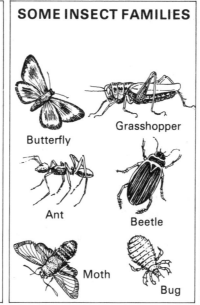

Butterfly
Grasshopper
Ant
Beetle
Moth
Bug

HOUSEFLY

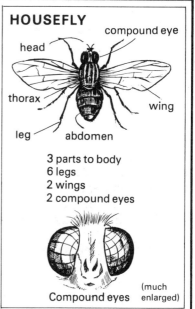

head · compound eye · thorax · wing · leg · abdomen

3 parts to body
6 legs
2 wings
2 compound eyes

Compound eyes (much enlarged)

METAMORPHOSIS IN INSECTS

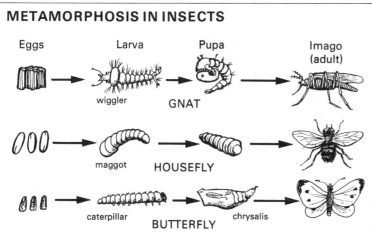

Eggs · Larva · Pupa · Imago (adult)

wiggler — GNAT

maggot — HOUSEFLY

caterpillar — chrysalis — BUTTERFLY

DRAGONFLY

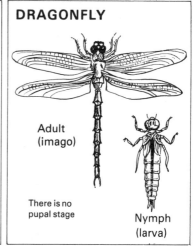

Adult (imago)

There is no pupal stage

Nymph (larva)

KEEPING FLIES

cotton wool
blowfly
meat
bottle

KEEPING SPIDERS

lid
jar
small air-holes
spider
webs
twig

SPIDERS

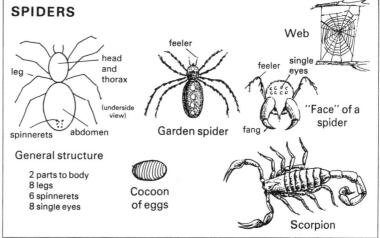

leg · head and thorax · (underside view) · spinnerets · abdomen

General structure

2 parts to body
8 legs
6 spinnerets
8 single eyes

feeler
Garden spider

feeler · single eyes · fang
"Face" of a spider

Web

Cocoon of eggs

Scorpion

Social insects

Insects which live together in colonies, such as bees, wasps and ants, are called *social insects*.

Bees

Bees often nest in holes in trees or *hives*. Hives are artificial nests made by men. The work in a hive is done by *workers;* these are females which do not lay eggs. All the eggs are laid by the *queen*. There is only one queen in a hive. Male bees are called *drones*.

Bees undergo metamorphosis. Their eggs are laid in a *comb*. This is made up of wax cells. Some of these are used for storing the honey and pollen on which the worker bees feed the larvae.

Wasps are closely related to bees in appearance and habits.

Ants

There are male, queen and worker ants. The workers are wingless female ants which do not lay eggs.

Winged male and female ants fly off together to start new colonies. The male dies after mating. The queen then starts a new colony of her own.

The white "ants' eggs" which you see in an ants' nest are not eggs at all. They are pupae. The pupae, which are covered in cocoons of a white, silky material, contain the larvae, or grubs. The larvae do not move about but they are not inactive. They are undergoing the metamorphosis from larvae into adult ants.

Ants live in a more complicated way than bees. Each worker has its own job to do. Some ants defend the nest, some build passages, some feed the larvae, and so on. Ants keep their own "cows". These are plant lice which secrete a sugary liquid that ants like. Ants introduce leaves into their nests. Then they eat the fungi which grow on these leaves.

Keeping ants

One of the diagrams shows you how to make a *formicarium,* or a home for ants. Make winding tunnels in the damp soil. Seal the edges of the plastic sheet and the wooden baseboard with tape. Transfer ants, eggs, larvae and pupae from an ants' nest into the food chamber. Keep the formicarium dark with a cloth cover. After a few days, you will notice that more tunnels have been built and that the eggs, larvae and pupae have been collected into different places. Feed the ants on honey, cake crumbs, dead insects, small pieces of fruit, etc.

Butterflies and moths

Butterflies and moths are similar insects, but there are some important differences between them. These differences are shown in the illustration opposite.

There are exceptions to these general characteristics. Some moths, for example, are very brightly coloured.

Keeping caterpillars

One of the diagrams shows how an insect cage can be made from a shoe box and a thin sheet of transparent plastic. Transfer some eggs and caterpillars from the garden to the box. Watch their development. The food should be changed and the cage cleaned out daily. Feed the caterpillars on leaves from the plants on which they were found. Put a little sawdust at the end of the box. This protects the pupae against coldness and damp.

Rearing silkworms

Silkworms are the caterpillars of the silk moth. Silk is obtained from the cocoons which they spin.

Place some silkworm eggs (these can be purchased) in an insect cage. Keep the cage in a warm place and the eggs will hatch. Feed the silkworms on mulberry or lettuce leaves. After a series of moults they will become quite large. Provide them with paper bags in which they can spin their cocoons. Transfer the cocoons to a dish of hot water. Wind the loosened silk on to a piece of cardboard.

BEES

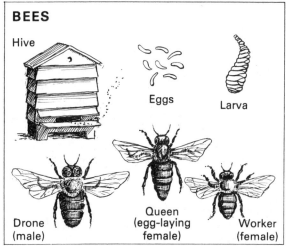

Hive

Eggs

Larva

Drone
(male)

Queen
(egg-laying
female)

Worker
(female)

KEEPING ANTS

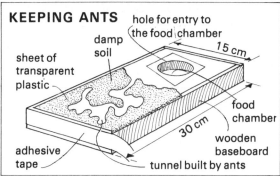

hole for entry to
the food chamber

damp
soil

15 cm

sheet of
transparent
plastic

food
chamber

30 cm

adhesive
tape

wooden
baseboard

tunnel built by ants

KEEPING CATERPILLARS

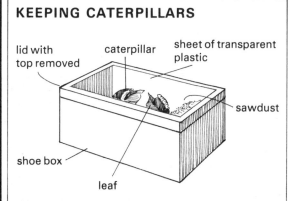

lid with
top removed

caterpillar

sheet of transparent
plastic

sawdust

shoe box

leaf

ANTS

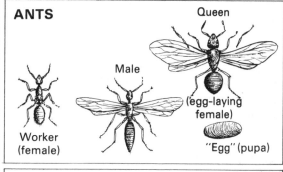

Queen

Male

Worker
(female)

(egg-laying
female)

"Egg" (pupa)

BUTTERFLIES AND MOTHS

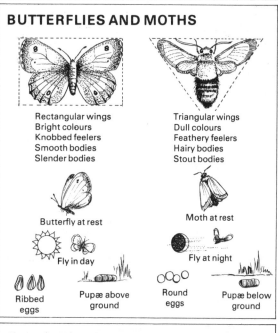

Rectangular wings
Bright colours
Knobbed feelers
Smooth bodies
Slender bodies

Triangular wings
Dull colours
Feathery feelers
Hairy bodies
Stout bodies

Butterfly at rest

Moth at rest

Fly in day

Fly at night

Ribbed
eggs

Pupæ above
ground

Round
eggs

Pupæ below
ground

CABBAGE WHITE BUTTERFLY

REARING SILKWORMS

STAGES

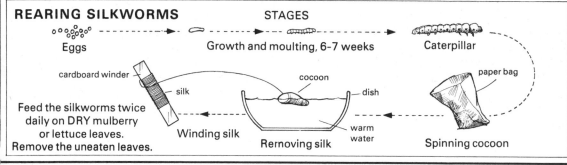

Eggs

Growth and moulting, 6-7 weeks

Caterpillar

cardboard winder

cocoon

paper bag

silk

dish

Feed the silkworms twice
daily on DRY mulberry
or lettuce leaves.
Remove the uneaten leaves.

Winding silk

Rernoving silk

warm
water

Spinning cocoon

Molluscs

A *mollusc* is an animal with a soft, simple body, a muscular creeping foot and a shell. There are exceptions. Some molluscs, such as mussels and oysters, are stationary. Octopuses and slugs do not possess shells.

Snails, whelks, limpets, oysters, cockles, winkles, mussels, octopuses, escallops and slugs are molluscs.

A snail lays pieces of jelly, about 6 mm wide and 12 mm long, each containing about 25 eggs. Young snails hatch out after a month. They take two years to become fully grown.

Collecting shells

Make a labelled collection of mollusc shells. A fishmonger will supply some of these. Examine and draw some of these shells.

A snail

Examine either a garden snail or a pond snail. Notice its shell, feelers, eyes, breathing hole and mouth. Make a labelled drawing.

Notice how it glides along on its foot. A garden snail leaves a slime track.

Pearls

Sometimes, a small grain of sand gets inside an oyster shell. The oyster protects itself against irritation by secreting layers of *calcium carbonate* around the sand grain to form a pearl. Small pearls are often found in mussels. Pearl oysters, which live in the Indian Ocean and the Pacific Ocean, produce valuable pearls.

Some simple animals

Jellyfishes, starfishes and sea anemones are simple animals found on the sea-shore.

The jellyfish feeds on other sea animals. Its *tentacles* are covered in stings with which it poisons its prey.

If a starfish loses one of its five limbs, it grows a new one. In fact, a lost limb can grow into a new starfish.

A sea anemone has the appearance of a beautiful flower when its tentacles are open. It catches food and defends itself with its tentacles. It is a stationary animal. It attaches itself firmly to a rock or the empty shell of a mollusc.

Some bath sponges are made from the soft sponges found in the Mediterranean Sea. The bodies of these animals are porous and elastic and will easily absorb large quantities of water.

Coral islands

Coral islands are formed from the accumulated skeletons of *coral polyps*. Many islands in the Pacific Ocean have been formed in this way. The secretions from these tiny creatures, which live in colonies, form cups of calcium carbonate. Coral polyps will only live in warm seas.

Worms

Worms are simple animals with long, narrow bodies. Earthworms, horseleeches and lugworms belong to this family.

Lugworms are found on the sea-shore and are used as bait by fishermen. Leeches live in water. They suck the blood of water animals.

The body of an earthworm is made up of segments. Its head is at the pointed end. It has a mouth but no eyes and ears. It breathes through its moist skin.

Very tiny animals

There are many tiny animals which are too small to be seen with the naked eye. One of these is the *amoeba*. It lives in the soil and mud at the bottom of ponds and ditches. This animal, when seen through a microscope, looks like an irregularly-shaped piece of colourless, transparent jelly. It moves along by expanding and contracting different parts of its soft body. There is a dark spot in its body that is called the *nucleus*. This is its *nervous system*.

An amoeba reproduces itself by dividing into two.

An amoeba

Examine a living amoeba under a microscope. The animal should be suspended in water on a slide. Notice its movements. Make a drawing.

MOLLUSCS

(not to scale)

Octopus   Snail   Slug   Oyster

Mussel   Whelk   Cockle   Winkle   Escallop

SNAILS

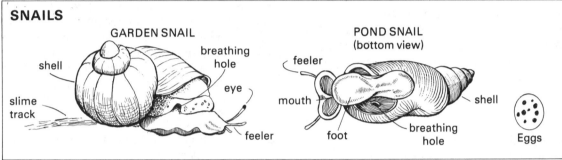

GARDEN SNAIL

breathing hole

shell

eye

slime track

feeler

POND SNAIL
(bottom view)

feeler

mouth

shell

foot

breathing hole

Eggs

SOME SIMPLE ANIMALS

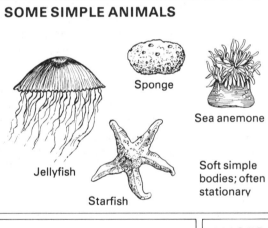

Sponge

Sea anemone

Jellyfish

Starfish

Soft simple bodies; often stationary

CORAL ISLANDS

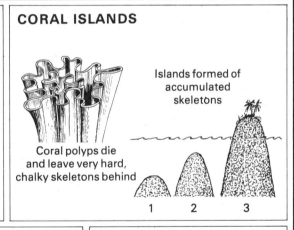

Islands formed of accumulated skeletons

Coral polyps die and leave very hard, chalky skeletons behind

1   2   3

WORMS

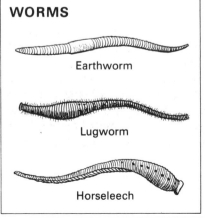

Earthworm

Lugworm

Horseleech

AMOEBA

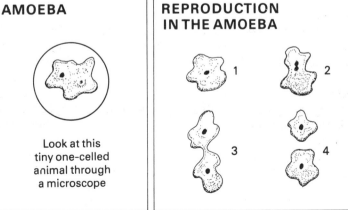

Look at this tiny one-celled animal through a microscope

REPRODUCTION IN THE AMOEBA

1

2

3

4

Fish

Fish live entirely in water. A fish has a streamlined body, and so it can move through water easily. Inside its body, there is an *air-bladder,* which gives it buoyancy. It has *pectoral* and *pelvic fins,* which correspond to our limbs, and *dorsal* and *tail fins.* Its body is covered in *scales.* It breathes by means of gills. The gills are protected by gill-covers.

The *hard roe* of a female fish consists of many eggs which are laid in masses in the water. *Milt* from the *soft roe* of the male fish is shed on top of the eggs in order to fertilize them. Many eggs are laid so that, despite the many dangers to which they are exposed, a few develop and become fully-grown fish.

British freshwater fish

Here are some of the freshwater fish in Britain:
Trout; gudgeon; stickleback; minnow; loach; bullhead; pike; roach; tench; carp; perch; bream; rudd; dace; chub.

Salmon and eels

Salmon and eels live in both salt and fresh water.

Salmon travel up rivers from the sea in order to *spawn,* that is, lay their eggs. The young fish return to the sea when they are two years old.

Eels travel for many miles to the Sargasso Sea to spawn. They die there. Some of the young eels, or *elvers,* as they are called, travel back to Europe. Others find their way to the East coast of North America.

A goldfish

Examine and make a labelled drawing of a goldfish. Notice its eyes, mouth, nostrils, gill-covers, scales, tail and paired fins.

A stickleback

Examine and make a labelled drawing of a stickleback. Notice its *spines.*

Other water animals

There are many water animals which are not vertebrates; therefore, they are not fishes.

Leeches, crayfish and newts are water animals that are not fishes. Newts, of course, are amphibians.

Crustaceans

There is a group of water animals, closely related to the insects, which have hard *exoskeletons,* or external skeletons. They are called *crustaceans.*

Crayfish, lobsters, crabs, woodlice, prawns, shrimps and water-fleas (*daphneae*) are crustaceans. They have jointed limbs for walking and swimming. They have feelers, and eyes on stalks. Some, such as crabs and lobsters, have strong claws. Crustaceans moult their hard shells as they grow. The woodlouse lives on land.

Observing freshwater animals

Here are some of the freshwater animals which you should observe. Notice how they feed, move, breathe and reproduce.

Minnow; stickleback; goldfish; freshwater shrimp; freshwater mussel; water-flea; water louse; crayfish; pond skater; water-boatman; horseleech; pond snail; carp; dragonfly nymph.

Miniature aquaria

Miniature aquaria can be made from large jam-jars. Put a layer of clean gravel in each jar. Place water plants in the gravel. Fill the jars with pond water. Keep the jars away from bright sunlight. Water animals can be kept separately in these jars. Do not keep different kinds of animals in the same aquarium, for many water animals are carnivorous. These jars are useful when living specimens are required for only a few days.

Sea animals

The following animals can be kept in sea-water:
Shrimp; prawn; mussel; sea snail; shore crab; cockle; sea anemone.

Feed them on small pieces of fish and chopped earthworms. A shallow dish is better than a jar because a large surface area of water is exposed to the atmosphere—the animals breathe the air dissolved in the water. Use seaweeds instead of freshwater plants.

Notice how the animals move, feed and breathe.

SOME BRITISH FRESHWATER FISH

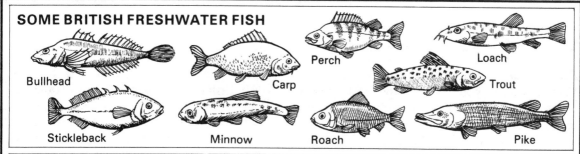

Bullhead

Perch

Loach

Carp

Trout

Stickleback

Minnow

Roach

Pike

LEAPING SALMON

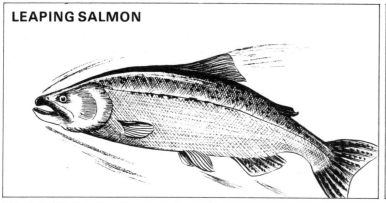

SEA-WATER AQUARIUM

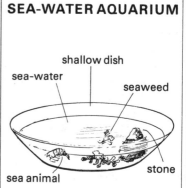

shallow dish

sea-water

seaweed

sea animal

stone

EEL AND ELVERS

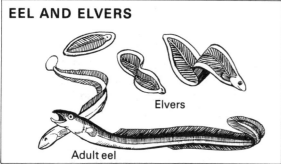

Elvers

Adult eel

GOLDFISH

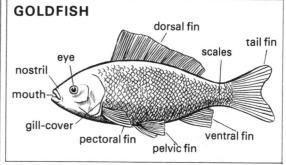

dorsal fin

eye

scales

tail fin

nostril

mouth

gill-cover

pectoral fin

pelvic fin

ventral fin

STICKLEBACK

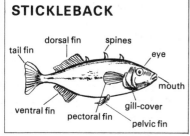

tail fin

dorsal fin

spines

eye

mouth

ventral fin

pectoral fin

gill-cover

pelvic fin

MINIATURE AQUARIUM

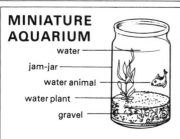

water

jam-jar

water animal

water plant

gravel

CRUSTACEANS

(not to scale)

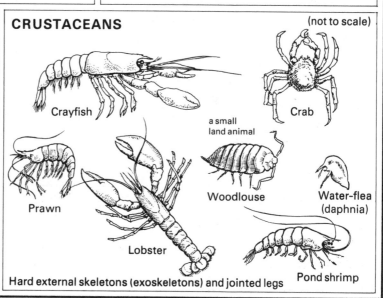

Crayfish

Crab

a small land animal

Prawn

Woodlouse

Water-flea (daphnia)

Lobster

Pond shrimp

Hard external skeletons (exoskeletons) and jointed legs

Pond dipping

Visit a pond in your locality and collect some water animals and plants. The only equipment required is a large meat-pan, a pond net, a spoon and a few jam-jars. One of the diagrams shows how a pond net can be made.

Collect insects, larvae, worms, plants, etc. Place them in the meat-pan and sort them. Use the spoon to transfer the sorted animals and plants to separate jars. Only the fishes need water continually. The other pond animals can be kept in wet weeds.

Your teacher and your classmates will help you to find out their names.

Your expeditions can be extended to streams, rivers and canals. *Danger! Take care when you are near deep water.*

Aquarium animals

Look for some of the water animals mentioned on page 76. Keep them in a large aquarium tank.

Water plants

Water plants can be allowed to grow in your aquarium tanks. These plants, as well as providing food for some of the animals, give off oxygen which dissolves in the aquarium water. Here are some of the plants which should be collected:

Water crowfoot; canadian pondweed; duckweed; frogbit; water milfoil; oblong pondweed.

Canadian pondweed grows quite rapidly and, therefore, it is very useful in aquarium tanks.

Keeping an aquarium tank

These simple rules will help you in keeping an aquarium tank properly:

1. Do not overfeed the fish. Uneaten food decays and fouls the water, causing the fish to be ill.
2. Now and again, as a change from dried foods, give the fish a live diet—water-fleas, small earthworms, maggots, etc.
3. Do not overcrowd the tank with too many fish. Fish come to the surface when the water does not contain enough oxygen. A good rule is "two centimetres of fish to 4 litres of water".
4. Do not keep fish together which are likely to fight. Large fish may eat small water animals. Large fish bully small fish.
5. Keep a few snails in the tank. They feed on algae and decaying food and so help to keep the water clean.
6. Do not change the water too often. Sudden temperature changes can cause chills and deaths.
7. Report deaths of fish, appearances of disease and other unusual occurrences to your teacher immediately.

Making a fish-pond

You could make a pond in the school garden. The water plants and animals in a pond have the natural conditions under which they thrive. Pond water contains most of the food which water animals require. Also, a large number of living specimens are available at all times of the year, and many plants and animals, particularly insects, find their way to the pond.

If the pond is balanced, that is, if there are enough plants and animals of the right kinds, there is no need for cleaning and feeding. The animals' droppings are food for the pond plants.

One of the diagrams shows a cross-section of a pond. A deep end for winter hibernation and a shallow end for summer warmth are provided. Plants are grown in a 5 cm layer of soil which is held in place by a 5 cm layer of gravel. The pond should be 1.5 m in diameter at least.

Provide the pond with a narrow overflow pipe, covered by a wire gauze to prevent the loss of small animals, and leading to a drain or a sump.

Goldfish, dace, carp, sticklebacks and minnows are suitable fish to keep. Freshwater shrimps and mussels, newts, amoebae, the larvae of insects, water-fleas, pond snails and other water animals can be kept, too.

GARDEN POND

POND DIPPING

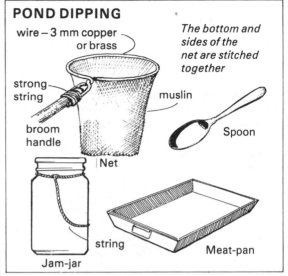

wire – 3 mm copper or brass

The bottom and sides of the net are stitched together

strong string

broom handle

muslin

Net

Spoon

string

Jam-jar

Meat-pan

SOME WATER PLANTS

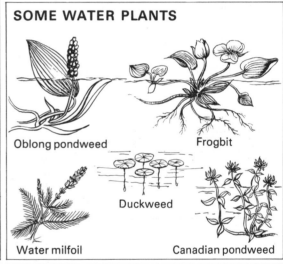

Oblong pondweed

Frogbit

Duckweed

Water milfoil

Canadian pondweed

MAKING A FISH-POND

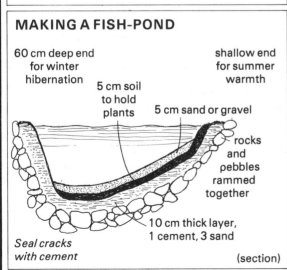

60 cm deep end for winter hibernation

shallow end for summer warmth

5 cm soil to hold plants

5 cm sand or gravel

rocks and pebbles rammed together

10 cm thick layer, 1 cement, 3 sand

Seal cracks with cement

(section)

POND OVERFLOW PIPE

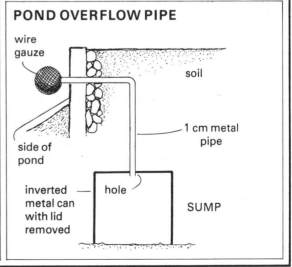

wire gauze

soil

side of pond

1 cm metal pipe

hole

inverted metal can with lid removed

SUMP

Exercises

1 Fruits and Seeds

1. Write down the names of *two* plants whose seeds are dispersed by (a) wind, (b) water, (c) animals, (d) birds, (e) special mechanisms.

2. Correctly pair these groups of words together to form proper sentences:

(a) Birds	are poisonous.
(b) Laburnum seeds	to prevent overcrowding.
(c) Seeds are dispersed	are carried by wind.
(d) Sycamore fruits	are fond of hips and haws.
(e) Coconuts	disperse their own seeds.
(f) Juicy fruits	float in water.
(g) Peas and beans	are carried in animal fur.
(h) Burdock fruits	are dispersed by birds.

2 Vegetative Reproduction

1. Write down the name of a plant which normally grows from (a) seeds, (b) bulbs, (c) corms, (d) tubers, (e) rhizomes, (f) runners.

2. Write down the names of *six* bulb-plants.

3. Complete these sentences. (Copy the sentences and fill in each blank space with a suitable word.)

(a) Tubers are swollen underground —————.
(b) ————— and artichokes are tubers. (c) Potatoes growing above the ground are ————— because sunlight causes the formation of chlorophyll. (d) The eyes of a potato are small —————. (e) Bulbs can be regarded as large buds with swollen food —————.
(f) The roots of a bulb are —————. (g) A ————— is a flat swollen stem filled with food. (h) A corm contains no leaves apart from the protective ————— leaves. (i) ————— are long underground stems which are sometimes swollen and contain food.

3 Preparing for Winter

1. Complete these sentences:

(a) During winter most plants have no leaves and ————— and transpiration is at a minimum. (b) Sap rises and plant growth is at a maximum during the ————— season. (c) Leaves fall during autumn and leaf scars are sealed off with ————— grown by the tree. (d) The age of a tree is shown by its number of ————— rings. (e) The distance between two ————— scars represents one year's growth of a twig. (f) ————— endings can be seen on the stalk of a fallen leaf. (g) The hard wood in the centre of a tree trunk is called —————. (h) The bud at the tip of a twig is called a ————— bud. (i) The breathing pores on a twig are called —————. (j) Small ————— buds on a twig grow and become lateral twigs.

2. Write sentences to show the meanings of these words. (The words must be contained in the sentences. Write a separate sentence for each word.)
Annual; sap; knot; herbaceous; terminal; girdle; deciduous; lateral.

4 Plant Growth

1. Complete these sentences:

(a) The growth movements of a plant are controlled by —————, light and gravity. (b) Plants grow towards sunlight; this is called —————. (c) Leaf tips are sensitive to —————. (d) The maximum growth in a root occurs behind the root —————. (e) *Geo* means "earth" and *hydro* means —————. (f) Plants grow toward water; this is called —————. (g) If horse-chestnut twigs are placed in water and kept warm their ————— open to display flowers and foliage.

5 Measuring Volume

1. Answer these questions (with full sentences):
(a) What are the units of length? (b) What are the units of volume? (c) Which measuring instrument is generally used to obtain an exact volume of water, say 50 cm³? (d) Which measuring instrument is used to obtain several volumes of water quickly, say 15 cm³, 23 cm³ and 19 cm³? (e) In the determination of a volume of sugar, a measuring jar and methylated spirit are used. Why is water not used as the liquid in this determination? (f) Why must a pipette not be used for measuring poisonous liquids?

2. Write sentences to show the meaning of these words:
Unit; capacity; meniscus; volume.

3. Some calculations:
(a) A burette reads 9 cm³. Volumes of 10 cm³ and 20 cm³ are delivered from it. What is the reading now? (b) A rectangular block of wood is 3 cm long, 2 cm wide and 1 cm high. What is its volume? (c) A measuring jar contains 30 cm³ of water. A pebble is put into the jar and the water rises to the 40 cm³ mark. What is the volume of the pebble? (d) A pebble sinker with a cork attached is put into a measuring jar containing 83 cm³ of water. The water rises to the level of the 94 cm³ mark. The pebble has a volume of 7 cm³. What is the volume of the cork?

Exercises

6 Measuring Mass and Weight

1. Answer these questions:

(a) Why is it better to measure apples by weight rather than by volume? (b) What is the weight of 1 litre of water? (c) What are the units of mass? (d) What are the units of weight?

2. Write a few sentences to explain the difference between mass and weight.

3. Some calculations:

(a) A spring balance indicates 100 g when an empty basket is suspended from its hook. Potatoes are placed in the basket. The balance now indicates 3.1 kg. What is the weight of the potatoes? (b) An empty litre measure has a weight of 100 gf. The measure, when filled with water, has a weight of 1.1 kgf. What is the weight of 1 litre of water? (c) Masses of 50 g, 10 g, 5 g, 1 g, 0.5 g and 0.05 g placed on the left-hand pan of a common balance are required to show the mass of the material placed on the right-hand pan. What is the mass of the material?

7 Density

1. Some calculations:

(a) 50 cm³ of a material weighs 25 gf. What is its (i) density, (ii) relative density? (b) What is the density of water, in kg/m³, if 2 m³ of water has a mass of 2000 kg? (c) A dish containing 50 cm³ of liquid weighs 30 gf. If the dish weighs 5 gf, what is the (i) density, (ii) relative density of the liquid? (d) What is the weight of 10 cm³ alcohol? The relative density of alcohol = 0.82. (e) What is the volume of 33.9 g of lead? The relative density of lead = 11.3. (f) A block of metal is 2 cm long, 1 cm wide and 0.5 cm high. It weighs 8 gf. What is the density of the metal? (g) 2 cm³ of alcohol weighs 1.64 gf. What is its (i) density, (ii) relative density? (h) How many times is silver heavier than cork? The relative density of silver = 10.5, and of cork = 0.25. (i) What is the weight of 10 cm³ of gold? The relative density of gold = 19.3.

2. Use an encyclopaedia to find out what you can about a *lactometer*.

8 Buoyancy

1. Answer these questions:

(a) Why does wood float on water? (b) Why do objects appear to weigh less in water than in air? (c) Why does iron sink in water? (d) Why do iron ships float? (e) Why does an egg float in salt water and sink in fresh water? (f) What are Plimsoll lines? (g) How is a submarine raised? (h) Why do bad eggs float? (i) What is meant by the letters L R on the side of a ship? (j) Why do swimmers float better in salt water than in fresh water?

2. Write down the Principle of Archimedes.

3. Make a simple drawing of the side of a ship to show the Plimsoll lines.

4. Here is a puzzle:

Two glass tumblers, one containing a cork, are completely filled with water. Which tumbler has the heavier contents?

5. Some calculations:

(a) A lump of metal weighs 14 gf. When suspended and immersed in water, its apparent weight is 4 gf. What is the weight of the water displaced by the metal? (b) A pebble which weighs 20 gf is suspended and immersed in water. The apparent weight of the pebble is 15 gf. What is the relative density of the pebble? (c) A piece of metal suspended and immersed in water has an apparent loss in weight of 5 gf. What is the weight of the metal? The relative density of the metal = 10. (d) A basin is placed in a trough which has been completely filled with water. The weight of the water which overflows is 100 gf. What is the weight of the basin? (e) A metal object of mass 2 g is suspended in water. 1 cm³ of water is displaced. What is the relative density of the metal? (f) An iron ship displaces 10 000 tonnes of water. What is the weight of the ship in kgf? (g) A barge displaces 200 m³ of water. What is the weight of the barge? 1 m³ of water weighs 1000 kgf.

9 Water Pressure

1. Answer these questions:

(a) Why is a diver's suit made of thick, reinforced material? (b) Why are the walls of reservoirs and dams thickest at their bases? (c) What is the effect of depth on water pressure? (d) For what purpose is the hydraulic press used? (e) How is an aeroplane undercarriage raised?

2. Make a labelled sectional drawing of a self-filling bucket.

3. Some calculations:

(a) 500 kg of water are contained in a cylinder which has a base area of 500 cm². Assume that 1 kgf = 10 N. What pressure, in N/cm², is exerted by the water on the base of the cylinder? (b) Water exerts a pressure of 0.1 N/cm² on a cylinder base which has an area of 5 cm². What mass of water is contained in the cylinder? (c) A can contains 20 kg of water. What is the force exerted on the bottom of the can?

Exercises

10 Air Pressure

1. Complete these sentences:
(a) Air can be compressed because it is —————.
(b) The ————— in a balloon presses in all directions.
(c) Air is used as a *cushion* in pneumatic —————.
(d) Soap bubbles are round in shape because air exerts the same ————— in all directions. (e) Otto von Guericke demonstrated the enormous pressure of the atmosphere with two metal —————. (f) A football bounces because the ————— which it contains acts as a *spring*. (g) Footballs and tennis balls keep their shapes because they contain air under —————. (h) The water in a dropping tube is held in place by atmospheric —————. (i) A space which contains no air is called a —————.

2. Write sentences to show the meanings of these words: Expel; compress; elastic; hemisphere; vacuum.

11 Using Air Pressure

1. Write down the names of *ten* common devices which are operated by air pressure.
2. State the purpose of the following:
(a) The small hole at the bottom of a drinking fountain.
(b) The hole in the lid of a toy sprinkler. (c) The washer in a bicycle pump.
3. Make labelled sectional drawings of (a) an animal drinking fountain, (b) a syringe, (c) an automatic flushing tank, (d) a bicycle pump, (e) a lift pump, (f) a force pump.

12 Measuring Air Pressure

1. Complete these sentences:
(a) Air in the atmosphere exerts a ————— of about 10 N/cm^2. (b) Air supports a ————— column about 10 m high. (c) A water column weighing ————— kgf and contained in a tube with a cross-sectional area of 1 cm^2 is just supported by atmospheric pressure. (d) Air pressure supports a ————— column about 76 cm high. (e) Air pressure falls 1 cm of ————— for every 120 m increase in altitude. (f) An ————— is an aneroid barometer marked in metres. (g) The vacuum at the top of a mercury barometer tube is named after the Italian scientist —————. (h) ————— means "without liquid". (i) A ————— is used for recording air pressures automatically. (j) A falling barometer indicates ————— weather.
2. Make a labelled sectional drawing of an aneroid barometer.

3. Some calculations:
(a) What is the force in newtons exerted by the atmosphere at a pressure of 100 000 Pa on a surface of 3 cm^2? (b) What is the air pressure in pascals if the air exerts a force of 204 000 N on an area of 2 m^2? (c) What air pressure reading, in cm of mercury, will be shown by a barometer at a height of 360 m? Atmospheric pressure at sea level is 76 cm of mercury.

13 Oxidation

1. Complete these sentences:
(a) Breathing, rusting and burning are forms of ————— (b) Carbon + ————— = carbon dioxide. (c) ————— + oxygen = iron oxide. (d) Rusting is the oxidation of —————. (e) Magnesium + oxygen = ————— oxide. (f) The oxidation of a metal causes an increase in its —————.
2. Write a few sentences to explain the *four* methods that are used to prevent rusting.
3. Write sentences to show the meanings of these words: Oxide; galvanize; bleaching; fumigant; litmus.

14 Acids and Alkalis

1. Complete these sentences:
(a) Litmus paper is ————— in colour in the presence of an acid. (b) Litmus paper is blue in colour in the presence of an —————. (c) Sulphuric acid, hydrochloric acid and nitric acid are called the ————— acids. (d) Acids taste —————. (e) ————— contain hydrogen. (f) ————— are often corrosive. (g) Acid + base = ————— + water. (h) Hydrochloric acid + ————— = table salt + water. (i) The poison in a bee's sting is —————. (j) All bases contain —————.
2. Write down the names of *four* materials which are (a) acids, (b) alkalis, (c) salts.
3. Write brief explanations of these statements:
(a) Caustic soda is alkaline. (b) Litmus is an indicator for acids and alkalis. (c) An acid is neutralized by a base. (d) Wasp stings should be treated with vinegar or the juice of a lemon.
4. Obtain pictures of common materials which are acids and alkalis. Pictures of acid fruits, bottles of vinegar and smelling salts, soap wrappers, etc., are suitable. Paste these, with suitable labels, into your notebook.

15 Acids and Hydrogen

1. Write down *three* uses of acids.
2. Write a few sentences about the weight, burning, colour, smell, preparation and uses of hydrogen.

Exercises

3. Answer these questions:
(a) What are the main ingredients of fruit salts? (b) What is the purpose of the citric acid in fruit salts? (c) What is the cause of effervescence when water is added to fruit salts? (d) What material is formed when hydrogen is burned in oxygen? (e) What happens when dilute hydrochloric acid is added to sodium bicarbonate?

16 Solutions
1. Complete these sentences:
(a) Salt dissolves in water to form a —————. (b) The liquid part of a solution is called the —————. (c) The solid which dissolves in a solvent is called the —————. (d) Wood is insoluble in —————. (e) ————— is soluble in water. (f) ————— feed on the mineral salts dissolved in soil water. (g) Mineral salts in the ground are ————— by rain-water and carried into the seas and oceans. (h) A solution in which no more of a solute will dissolve is said to be —————. (i) Photographers' *hypo* is very —————. (j) The solubility of a substance usually increases with a rise in —————. (k) A mixture of iron filings and salt could be separated by —————. (l) The deposit left on a filter paper is called the —————.
2. Write sentences to show the meanings of these words:
Solution; soluble; solvent; solute; saturated; filtrate.
3. Suggest solvents for these materials:
Sugar; table salt; shellac; lime; iodine; oil; alum; grease; Epsom salt; *hypo*.

17 Crystals
1. Correctly pair these groups of words together to form proper sentences:

(a) Crystals of the same substance	are double pyramids.
(b) Common salt crystals	crystalline or amorphous.
(c) Alum crystals	do not contain water.
(d) All substances are either	water from the atmosphere.
(e) Glass is an	is deliquescent.
(f) Anhydrous salts	have the same shape.
(g) Deliquescent salts absorb	are rectangular blocks.
(h) Table salt	amorphous substance.
(i) Glauber's salt is	blue when moistened.
(j) Anhydrous copper sulphate turns	efflorescent.

2. Write sentences to show the meanings of these words:
Amorphous; anhydrous; deliquescent; efflorescent; suspension; stalactite; stalagmite.

18 Pure Water
1. Answer these questions:
(a) Why is sea-water not used for drinking? (b) Why are fluoride salts in drinking water beneficial? (c) Why do campers boil water from rivers before drinking it? (d) Why is fresh water better for washing than sea-water? (e) Why are Spa waters believed to be helpful to rheumatic sufferers? (f) Why do people sometimes put marbles and pebbles in kettles? (g) Why is soft water used in steam-engines? (h) What is distilled water? (i) Why are small quantities of potassium permanganate sometimes added to water from rivers? (j) Why is it easy to make a good lather with rain-water?
2. Make a list of the impurities which might be found in a bucket of river water.
3. Write a short essay with the title *At the Waterworks*. Expand these paragraph notes. Include labelled drawings.
Reservoirs. Rivers and lakes to reservoirs. Pumping station. Rubbish sinks. Germs killed by oxygen in the atmosphere.
Filter beds. Filters are layers of gravel and sand. Suspended materials caught in filters. Some germs filtered out by layer of green algae. Inlet pipes point upwards to prevent disturbance of sand.
Chlorination. Chlorine kills germs. Chlorine is poisonous to humans. Small quantities only are used.
Water-towers. Water stored ready for consumers. No pumping to taps required.
4. Visit a waterworks. Make notes and sketches about some of the things you see.
5. Write a few sentences to explain how you would remove the fur from the inside of a kettle.

19 Surface Tension
1. Rearrange these groups of jumbled words to make sentences:
(a) a tap from drops water in shape round are (b) surface the invisible *skin* surface water is on called tension the of (c) possible water is it to float needles on (d) heavier steel than water is (e) the tension surface oil of is than less water of (f) carry disease the mosquitoes malaria (g) some a pond surface insects and walk run the on of (h) swamps mosquitoes sprayed often lakes and where are breed oil with (i) soda emulsifying is caustic an agent

Exercises

2. Write sentences to show the meaning of these words: Tension; taut; spherical; larvae; emulsion.

3. Make simple drawings of (a) a whirligig beetle, (b) a pond skater, (c) a mosquito larva.

20 Capillary Attraction

1. Write a few sentences to explain what is meant by capillary attraction.

2. Write down the names of *five* common devices which depend upon capillarity for their operation.

3. Make simple labelled drawings of (a) a model spirit-burner, (b) a petrol lighter.

4. Write brief explanations of these statements:

(a) Water birds have natural oil on their feathers. (b) Builders provide houses with damp courses. (c) Umbrellas and oilskins are made of non-porous materials. (d) Outside woodwork is often creosoted. (e) A roof is erected at an angle and so helps to keep a building waterproof.

21 Washing and Cleaning

1. Complete these sentences:

(a) Oil and dirt carry ——————. (b) Germs cause ——————. (c) —————— salts are made of sodium carbonate, dye and perfume. (d) Sweat sometimes smells ——————. (e) —————— is a water-softener. (f) Wood —————— from beneath cooking pots was once used for cleaning. (g) Tallow is —————— fat. (h) Scouring soaps contain an abrasive, such as —————— or powdered pumice. (i) Ammonia is a —————— solvent. (j) Soap powders contain a water-softener, such as sodium carbonate or ——————.

2. Write a few sentences to explain how you would remove a tar stain from your coat.

22 Heat Movement

1. Answer these questions:

(a) What are the three ways in which heat moves? (b) Is warm water heavier than cold water? (c) Which metal is liquid at ordinary temperatures? (d) How do heat and light move across space? (e) Why do metal objects feel colder than non-metal objects? (f) Why are kettles and pans usually made of aluminium? (g) Is water a good heat conductor? (h) Who invented the miner's safety lamp? (i) How are coal mines lit today? (j) For what purpose is the safety lamp still used?

2. Make a drawing of a miner's safety lamp.

3. Make simple drawings to show what is meant by conduction, convection and radiation.

23 Convection

1. Correctly pair these groups of words together to form proper sentences:

(a) When water is heated,	by convection currents.
(b) Heat moves in liquids	of central heating.
(c) Dust particles are carried upwards	in ventilation.
(d) Winds are caused	it expands.
(e) A fire with a tall	by currents of warm air.
(f) Convection currents help	by convection.
(g) The Romans had a form	ocean currents.
(h) Heat from the Sun causes	chimney burns well.
(i) Fires were once used	than an equal volume of warm water.
(j) A volume of cold water is heavier	to ventilate coal mines.

2. Make labelled drawings to show how convection currents are used in chimneys and in the ventilation of rooms and mines.

24 Radiation

1. Complete these sentences:

(a) —————— waves are radiated by a lamp. (b) The Sun radiates —————— and light. (c) Dark, unpolished surfaces absorb heat ——————. (d) Bright, polished surfaces absorb heat ——————. (e) In tropical climates people sometimes wear —————— clothing which helps them to keep cool. (f) —————— soils absorb more heat than light soils. (g) Heat from the Sun reaches the Earth by ——————. (h) The atmosphere is thin at great ——————.

2. Write brief explanations of these statements:

(a) Gardeners sometimes cover the soil with black polythene sheeting. (b) Greenhouses and garden frames are covered with glass. (c) The reflectors in electric fires have "silvered" surfaces.

3. Make a labelled sectional drawing of a vacuum flask.

Exercises

25 The Sky at Night

1. Write sentences to show the meanings of these words:
Star; prophecy; constellation; hemisphere; latitude; orbit; incandescent; comet; meteor; meteorite.

2. Write a short essay with the title *The Night Sky*. Make your own paragraph notes in the manner shown in a previous exercise.

3. Make labelled drawings of (a) the Southern Cross, (b) the Plough, (c) the Constellation Cygnus.

4. Write a few sentences about *Halley's Comet*. Use an encyclopaedia.

26 The Solar System

1. Complete these sentences:
(a) The Sun and its planets are known as the —————— system. (b) These planets revolve around the Sun in regular ——————. (c) The planets are ——————, Venus, Earth, ——————, Jupiter, Saturn, Uranus, Neptune and Pluto. (d) Between —————— and Jupiter there is a ring of nearly 44 000 minor planets called ——————. (e) The largest planets are —————— and Saturn. (f) The smallest planets are —————— and Mercury. (g) Saturn has —————— moons. (h) Mercury is too —————— for life. (i) —————— and Pluto are too cold for life. (j) Some of the —————— have little or no atmosphere. (k) The planets are made visible to us by —————— reflected from them. (l) —————— is sometimes called the Morning Star. (m) The Sun appears to move from east to ——————. (n) The distance between the Sun and the Earth is about —————— million kilometres. (o) The side of the Earth in —————— is having night. (p) The Moon is —————— thousand kilometres away from the Earth. (q) There is no —————— on the Moon. (r) The surface of the Moon is covered in volcanic ——————. (s) The Moon's surface is extremely hot during its day and extremely —————— at night.

2. Write sentences to show the meanings of these words:
Planet; asteroid; lunar; phase; eclipse.

3. Some calculations:
(a) How long does light take to travel 600 000 km? (b) On the model solar system described on page 58, the distance between the Sun and Mercury is given as 60 cm. What is the actual distance between the Sun and Mercury? (c) What time, in days, is taken by the Moon in making two revolutions around the Earth?

27 Gravity

1. Write a few sentences about each of these topics:
(a) Force of gravity. (b) Centrifugal and centripetal force. (c) Tides. (d) Centre of gravity. (e) Stability. (f) Artificial satellites.

2. Write brief explanations of these statements:
(a) Planets are not drawn into the Sun by the force of the Sun's gravity. (b) The string of a plumb-line is vertical. (c) A double-decker bus does not fall over when travelling on the curved surface of a road.

3. Make a drawing of a plumb-line.

28 Measuring Time

1. Answer these questions:
(a) How can time be defined? (b) How is time measured? (c) How can you use your own shadow as a clock? (d) Where is Cleopatra's Needle? (e) At what time of the day is the shadow of a stick shortest? (f) In which direction do shadows point at midday in the Northern Hemisphere? (g) Why can a sundial be used only when the Sun is shining? (h) What is a solar year? (i) Which is the longest day in the year? (j) What is the period of a pendulum? (k) What is a solar day? (l) Where are pendulums used?

2. Write a few sentences about each of these topics:
(a) Cleopatra's Needle. (b) Earth's rotation. (c) Seasons. (d) Pendulum.

3. Make drawings of some of the devices used for measuring time.

29 The Animal Kingdom

1. Make *two* separate lists of vertebrate and invertebrate animals. Choose your animals from this list:
Dog; frog; toad; earthworm; sparrow; rabbit; oyster; blackbird; amoeba; eagle; herring; plaice; mouse; newt; lizard; eel; turtle; stickleback; gnat; jellyfish; wasp; adder; swallow; whale; duck; tortoise; gorilla; salmon; garden-spider; mussel; earwig; slug; starfish; crocodile; octopus; leech; sponge.

2. Under the headings *mammals, birds, reptiles, amphibians* and *fishes*, make *five* separate lists of the vertebrate animals given in question 1.

3. Write brief explanations of the ways in which these animals are adapted to the conditions under which they live:
(a) Giraffe. (b) Herring. (c) Frog.

Exercises

4. Complete these sentences:
(a) Mammals and ——————— are warm-blooded.
(b) Birds lay eggs but ——————— do not. (c) Mammals have fur, hair or wool and birds have ———————.
(d) Mammals and birds breathe by means of ———————.
(e) Reptiles, ——————— and fishes are cold-blooded.
(f) Fishes and ——————— are covered in scales.
(g) Amphibians live on ——————— and in water. (h) The eggs of ——————— are shell-less. (i) Fishes breathe by means of ———————. (j) Amphibians have both gills and ——————— at some stages in their lives. (k) Tadpoles breathe with ——————— in the early stages of their lives.
(1) Fishes lay many ———————. (m) Fishes move by means of ———————. (n) Some reptiles do not lay ———————. (o) Mammals supply their ——————— with milk.
5. Write down the names of *two* animals which rely for defence upon (a) speed, (b) size, (c) claws, (d) teeth, (e) poison, (f) horns, (g) beaks, (h) camouflage, (i) smell, (j) sight, (k) hearing.
6. Write down the names of *eight* animals which hibernate during the winter.
7. What special names are given to the homes of (a) birds, (b) squirrels, (c) rabbits, (d) foxes, (e) badgers, (f) hares, (g) garden snails?

30 Mammals and Birds
1. Write down the names of *two* mammals which are (a) pouched, (b) insect-eating, (c) gnawing, (d) marine, (e) hoofed, (f) flesh-eating, (g) manlike.
2. Make a list of *ten* mammals which live wild in Britain.
3. Rearrange these groups of jumbled words to make sentences:
(a) mammals on carnivores feed flesh which called are
(b) herbivores which on mammals plants are called feed
(c) mammals bats fly the are only which can (d) the bats of wings are membranes large skin of (e) horses dogs are domesticated and animals called (f) ducks swimming feet have webbed for
4. Make drawings of (a) the beaks of a heron, parrot, eagle and duck, (b) the feet of an eagle, pigeon, duck and woodpecker.
5. Make a list of *fifteen* birds which live wild in Britain.

31 Cold-blooded Animals
1. Write down the names of *three* British (a) reptiles, (b) amphibians.

2. Write down the names of *ten* reptiles.
3. Make labelled drawings to show the main stages in the life history of a frog.
4. Write sentences to show the meanings of these words: Amphibian; metamorphosis; spawn; tadpole; vivarium.

32 Insects and Spiders
1. Complete these sentences:
(a) A housefly has two compound eyes and ——————— legs. (b) The body of an ——————— is in three parts—the ———————, the thorax and the abdomen. (c) The third pair of legs of a grasshopper is used for ———————.
(d) The front pair of wings of a ——————— acts as a hard, protective cover. (e) Insects which have mouth-parts for sucking juices are called ———————. (f) Spiders are not ———————. (g) There are two parts to the body of a ———————. (h) Spiders use their ——————— for trapping insects. (i) Spiders wrap their eggs in protective ——————— of silk. (j) A garden-spider has eight legs and ——————— eyes. (k) There are ——————— spinnerets under the abdomen of a house-spider. (l) Spiders do not undergo ———————; baby spiders look the same as adult spiders. (m) ——————— and king-crabs belong to the spider family. (n) Insect larvae lose their skins by ———————. (o) Spiders ——————— their victims with their fangs.
2. Write sentences to show the meanings of these words: Thorax; abdomen; compound; bug; larva; pupa; imago; nymph; spinneret; cocoon; arachnid.
3. Make labelled drawings to show the main stages in the life-history of (a) a butterfly, (b) a gnat.
4. Make labelled drawings to show the general structure of (a) an insect, (b) a spider.

33 Some Insect Families
1. Complete these sentences:
(a) Insects which live together in colonies are called ——————— insects. (b) Ants, ——————— and wasps usually live in colonies. (c) Female ants and bees are called ———————. (d) Male bees are called ———————.
(e) Eggs are laid by queen ants and queen ———————.
(f) Bee larvae live on pollen and ——————— provided by the workers. (g) Bees nest in trees or ———————.
(h) There is only one ——————— in a hive. (i) Worker ants are wingless ———————. (j) White "ants eggs" are really ———————. (k) Ants keep their own ———————; these are plant lice which secrete a sugary liquid.

Exercises

(l) Winged male and female ants fly off together to start new —————. (m) Each worker ————— has its own job to do. (n) Silkworms are ————— of the silk moth. (o) Silkworms feed on ————— or lettuce leaves.

2. Make a table to show the main differences between butterflies and moths.

34 Some Simple Animals

1. Complete these sentences:
(a) A ————— breathes through a single breathing hole. (b) Slugs and ————— do not possess shells. (c) A ————— glides along on its muscular foot. (d) Pearls are found in oysters in the Indian and ————— Oceans. (e) A snail uses its ————— as feelers and eyes. (f) A pearl consists mainly of calcium —————. (g) Bath sponges absorb —————. (h) An earthworm has no eyes and ears and its body is made up of —————. (i) A sea anemone has the appearance of a beautiful —————. (j) A ————— can grow new limbs.

2. Make a list of *eight* molluscs.
3. Make labelled drawings to show how islands have been formed from córal polyps.
4. Write sentences to show the meanings of these words: Stationary; tentacles; segments; colonies; nucleus.

35 Water Animals

1. Consider a goldfish and then state the purposes of these organs:
(a) Fins. (b) Tail. (c) Scales. (d) Gill-cover. (e) Air-bladder.
2. Make a list of *ten* British freshwater fishes.
3. Answer these questions:
(a) What is hard roe? (b) Why do fish lay many eggs? (c) Where do salmon spawn? (d) Where do eels spawn? (e) What are young eels called? (f) What is an exoskeleton? (g) How do crustaceans grow despite their hard shells?
4. Write down the names of *five* water animals which are not true fishes.
5. Write down the names of *six* crustaceans.

36 Pond Life

1. Complete these sentences:
(a) Pond animals can be kept in ————— weeds. (b) ————— need water continually. (c) Water plants provide food and ————— for water animals. (d) Canadian ————— grows quickly. (e) ————— should not be overfed. (f) Uneaten ————— in an aquarium goes bad and fouls the water. (g) Fish need a live diet as a change from ————— foods. (h) Fish swim to the top when their ————— does not contain enough oxygen. (i) All kinds of ————— should not be kept together. (j) Snails feed on —————. (k) Aquarium ————— should not be changed often. (l) Fish are very susceptible to —————. (m) Pond water contains most of the ————— which water animals require. (n) If a pond is well balanced with plants and ————— there is no need for cleaning and feeding.

2. Write down the names of *four* common water plants.
3. Suppose that you notice the following defects in an aquarium tank. Write a few sentences to explain how you would remedy each defect.
(a) Unpleasant smell and cloudy water. (b) Fish coming to the surface. (c) Excessive growth of algae. (d) Dead fish floating on the surface of the water.
4. Make a list of water animals which could be kept in a school garden pond.
5. Make a labelled sectional drawing of a fish pond suitable for the school garden. Make a sectional drawing of the overflow pipe also.

Answers to Numerical Exercises

Chapter 5: 3. (a) 39 cm³, (b) 6 cm³, (c) 10 cm³, (d) 4 cm³.
Chapter 6: 3. (a) 3 kgf, (b) 1 kgf, (c) 66.55 g.
Chapter 7: 1. (a) (i) 0.5 g/cm³, (ii) 0.5, (b) 1000 kg/m³, (c) (i) 0.5 g/cm³, (ii) 0.5, (d) 8.2 gf, (e) 3 cm³, (f) 8 g/cm³, (g) (i) 8.2 g/cm³, (ii) 8.2, (h) 42, (i) 193 g.
Chapter 8: 5. (a) 10 gf, (b) 4, (c) 50 gf, (d) 100 gf, (e) 2, (f) 10 million kgf, (g) 200 000 kgf.
Chapter 9: 3. (a) 10 N/cm², (b) 50 g, (c) 20 kgf or 200 N.
Chapter 12: 3. (a) 30 N, (b) 102 000 Pa, (c) 73 cm.
Chapter 26: 3. (a) 2 s, (b) 60 million km, (c) 58 days.

Index